NO PANIC DIGITAL SAT [2025 Edition]:

Achieve a 1550+ Score Effortlessly with the Target Method™. Includes Video Tutorials, Secret Tactics, and Tips to Unlock Your Academic Career [98% Success Rate]

MindPress Publications

First Edition: 2024

TABLE OF CONTENTS

Introduction

Welcome to Your SAT Journey

Welcome to your ultimate guide for conquering the SAT! Whether you're just beginning your preparation or looking to refine your skills before the big day, this book is designed to be your companion every step of the way. The SAT is a crucial milestone in your academic journey, and performing well on this test can open doors to college admissions, scholarships, and other opportunities that can shape your future. Our goal is to provide you with the tools, strategies, and confidence you need to achieve your best possible score.

Purpose of the SAT

The SAT, or Scholastic Assessment Test, is a standardized test widely used for college admissions in the United States. It assesses your readiness for college by measuring skills in reading, writing, and mathematics—areas that are considered essential for academic success. The test is designed to evaluate your ability to think critically and solve problems, both of which are vital for college and beyond.

Understanding the significance of the SAT is the first step in your preparation. Colleges and universities use SAT scores as part of their admissions process to gauge your academic potential. A high SAT score can not only increase your chances of getting into your dream college but also make you eligible for scholarships and other forms of financial aid.

Overview of the Book

This book is structured to guide you through every aspect of the SAT, from understanding the test format to mastering each section. We've divided the book into several key sections:

1. **Diagnostic Test**: Start with a full-length diagnostic test to assess your current level. This will help you identify your strengths and weaknesses, allowing you to tailor your study plan accordingly.
2. **Math**: We'll break down the math section into its core components, providing in-depth explanations and practice problems. You'll learn everything from algebra and geometry to advanced math concepts, with tips on how to tackle even the trickiest questions.
3. **Evidence-Based Reading and Writing**: This section will help you sharpen your reading comprehension and grammar skills. We'll provide strategies for analyzing passages, understanding complex texts, and improving your writing mechanics.
4. **Essay (Optional)**: If your target colleges require or recommend the SAT essay, this section will guide you through crafting a high-scoring essay. We'll cover everything from brainstorming and planning to writing and revising.
5. **Test-Taking Strategies**: Beyond content mastery, test-taking strategies are crucial for success. This section will teach you how to manage your time, eliminate incorrect answers, and reduce test anxiety.
6. **Practice Tests**: Throughout the book, you'll find full-length practice tests that simulate the real SAT experience. These tests are designed to help you apply what you've learned and measure your progress.
7. **Motivational and Inspirational Resources**: Preparing for the SAT can be stressful, but staying motivated is key. This section includes success stories, inspirational quotes, and goal-setting tools to keep you on track.
8. **Supplementary Materials**: To complement your studies, we've included study schedules, a glossary of SAT terms, and other helpful resources. These tools will aid in organizing your preparation and deepening your understanding of key concepts.
9. **Bonus Materials**: To make this book the most comprehensive SAT prep resource available, we've added bonuses such as online resources access. These extras are designed to enhance your learning experience and give you an edge in your preparation.

How to Use This Book

This book is more than just a study guide; it's a roadmap to your success on the SAT. Here's how to make the most of it:

1. **Start with the Diagnostic Test**: Before diving into the content, take the diagnostic test to get a clear picture of where you stand. Use the results to focus your studies on areas that need improvement.
2. **Follow the Structured Sections**: Each section of this book is designed to build your skills progressively. Start with the basics and move on to more complex topics as you gain confidence.
3. **Practice Regularly**: The practice problems and tests in this book are essential for reinforcing what you've learned. Make sure to complete them as you go along, and review your mistakes to understand where you need more practice.
4. **Use the Test-Taking Strategies**: Mastering the content is only part of the battle. Use the test-taking strategies outlined in this book to approach the SAT with confidence and maximize your score.
5. **Stay Motivated**: Preparing for the SAT can be a long and challenging process. Use the motivational resources in this book to keep yourself inspired and on track.
6. **Make Use of the Bonus Materials**: Take advantage of the additional resources, including the online portal and personalized study plans. These tools are designed to provide you with extra support and enhance your learning experience.

Creating a Study Plan

A well-structured study plan is critical for SAT success. Without a plan, it's easy to feel overwhelmed or fall behind in your preparation. Here's a guide to creating an effective study plan:

1. **Set Your Goal**: Start by identifying your target SAT score. Research the average scores for the colleges you're interested in and set a goal that aligns with your aspirations.
2. **Assess Your Starting Point**: Use the diagnostic test to determine your baseline score. This will help you understand how much you need to improve and which areas require the most focus.
3. **Allocate Study Time**: Depending on how much time you have before the test, divide your study sessions into manageable chunks. We recommend dedicating specific days to different sections of the SAT to ensure balanced preparation.
4. **Prioritize Weak Areas**: Focus more time on the sections where you scored lower in the diagnostic test. However, don't neglect your stronger areas—consistent practice is key to maintaining and improving your skills.
5. **Include Regular Practice Tests**: Schedule full-length practice tests at regular intervals to monitor your progress. These tests will help you build stamina and get used to the timing and format of the SAT.
6. **Adjust Your Plan as Needed**: As you progress, you may find that some areas require more attention than others. Be flexible and adjust your study plan to meet your evolving needs.
7. **Stay Consistent**: Consistency is crucial when preparing for the SAT. Stick to your study schedule as closely as possible, and make studying a regular part of your routine.
8. **Rest and Recharge**: Don't forget to include breaks in your study plan. Rest is important for maintaining focus and preventing burnout.

The SAT is a significant challenge, but with the right preparation, you can conquer it. This book is your comprehensive guide to mastering the SAT, providing you with everything you need to succeed. From the moment you pick it up to the day you sit for the test, this book will be your partner in achieving your academic goals. Remember, success on the SAT isn't just about intelligence—it's about preparation, strategy, and perseverance. Let's embark on this journey together, and by the end, you'll be ready to achieve the score you've always dreamed of.

Diagnostic Test

Introduction to the Diagnostic Test

Before diving into the comprehensive study material this book provides, it's essential to understand where you currently stand in terms of your SAT readiness. The purpose of the diagnostic test is to give you a clear picture of your strengths and weaknesses. By taking this test under conditions that simulate the actual SAT, you'll gain valuable insights into your current performance level. This information will be crucial in creating a personalized study plan that focuses on improving the areas where you need the most help.

The diagnostic test is structured to mirror the real SAT as closely as possible. It includes sections on Math, Evidence-Based Reading, and Writing, as well as an optional Essay section. Taking this test seriously will provide you with a baseline score that can guide your preparation and help you track your progress as you work through the rest of the book.

How to Take the Diagnostic Test

To get the most accurate assessment from your diagnostic test, it's important to replicate the test conditions of the actual SAT as closely as possible. Here are some steps to follow:

1. **Set Aside Enough Time**: The SAT is a lengthy exam, so make sure you have enough uninterrupted time to complete the diagnostic test in one sitting. The test will take approximately 3 hours without the essay, and 3 hours and 50 minutes with the essay.
2. **Create a Quiet Environment**: Find a quiet, comfortable place where you won't be disturbed. This will help you focus and simulate the testing environment.
3. **Use Proper Timing**: Time each section according to the SAT's guidelines. Use a timer or stopwatch to ensure that you're adhering to the time limits. The actual timing for each section is as follows:
    - **Reading**: 65 minutes
    - **Writing and Language**: 35 minutes
    - **Math (No Calculator)**: 25 minutes
    - **Math (Calculator)**: 55 minutes
    - **Essay (Optional)**: 50 minutes

4. **Follow Test Protocols**: During the test, avoid using any resources or tools that wouldn't be allowed in the actual exam, such as notes, books, or unauthorized calculators. Stick to the allowed tools to get a realistic assessment of your abilities.
5. **Take Short Breaks**: Just like the real SAT, you can take short breaks between sections. Typically, the SAT allows a 10-minute break after the Reading section and a 5-minute break after the Math (No Calculator) section.
6. **Review Instructions Carefully**: Make sure you understand the instructions for each section before you begin. This will help you avoid simple mistakes and manage your time effectively.

Diagnostic Test Structure

The diagnostic test is divided into the same sections as the SAT:

1. **Reading**:
   - **Content**: This section includes passages from literature, historical documents, social sciences, and natural sciences. You'll be asked to answer multiple-choice questions that assess your ability to understand, interpret, and analyze these passages.
   - **Number of Questions**: 52 questions
   - **Time**: 65 minutes
2. **Writing and Language**:
   - **Content**: This section focuses on grammar, punctuation, and effective language use. You'll be asked to revise and edit passages to improve clarity, coherence, and correctness.
   - **Number of Questions**: 44 questions
   - **Time**: 35 minutes
3. **Math (No Calculator)**:
   - **Content**: This section tests your mathematical reasoning and problem-solving abilities without the use of a calculator. It includes questions on algebra, geometry, and some trigonometry.
   - **Number of Questions**: 20 questions
   - **Time**: 25 minutes
4. **Math (Calculator)**:
   - **Content**: This section allows you to use a calculator and covers a broader range of math topics, including more advanced algebra and data analysis.
   - **Number of Questions**: 38 questions
   - **Time**: 55 minutes
5. **Essay (Optional)**:
   - **Content**: If you choose to take the essay section, you'll be asked to read a passage and write an essay analyzing the author's argument. This section tests your reading comprehension, analytical, and writing skills.
   - **Time**: 50 minutes

Scoring the Diagnostic Test

After completing the diagnostic test, it's time to score it. Scoring the test accurately will give you a clear understanding of your current standing and help you identify areas that need improvement.

1. **Use the Answer Key**: Refer to the provided answer key to score your test. For each correct answer, give yourself one point. There is no penalty for incorrect answers, so simply tally up the correct responses.
2. **Calculate Your Raw Scores**: Your raw score for each section is the number of questions you answered correctly. There is no penalty for guessing, so even incorrect answers don't reduce your raw score.
3. **Convert Raw Scores to Scaled Scores**: The SAT uses a scaled score system, with each section scored between 200 and 800. Use the conversion tables provided in this book to convert your raw scores into scaled scores.
4. **Essay Scoring**: If you completed the essay section, use the provided rubric to score your essay. The essay is scored separately from the rest of the SAT and is reported on a scale of 2 to 8 in three areas: Reading, Analysis, and Writing.
5. **Calculate Your Total Score**: Add your scaled scores from the Math and Evidence-Based Reading and Writing sections to get your total SAT score, which ranges from 400 to 1600.

Analyzing Your Results

Once you've scored your diagnostic test, it's important to take the time to analyze the results. Understanding your performance in each section will help you identify where to focus your study efforts.

1. **Identify Strengths and Weaknesses**: Look at the individual section scores to determine where you performed well and where you need improvement. For example, if your Reading score is strong but your Math score is lower, you'll want to spend more time on math practice.
2. **Review Your Mistakes**: Go through the questions you got wrong and try to understand why. Was it a lack of knowledge, a misunderstanding of the question, or a careless mistake? This will help you avoid similar errors in the future.
3. **Set Target Scores**: Based on your diagnostic test results, set realistic target scores for each section. These targets should be challenging but achievable with diligent practice.
4. **Adjust Your Study Plan**: Use the insights from your diagnostic test to tailor your study plan. Allocate more time to areas where you need improvement and less time to areas where you're already strong.
5. **Track Progress**: Keep a record of your diagnostic test scores and compare them with future practice tests. Tracking your progress over time will help you stay motivated and focused on your goals.

The diagnostic test is a crucial first step in your SAT preparation journey. By taking this test seriously and analyzing your results, you'll gain valuable insights into your current abilities and set the stage for targeted, effective study. Remember, the goal of the diagnostic test is not to discourage you but to give you a clear starting point. With the right preparation and effort, you can significantly improve your scores and achieve your academic goals.

As you move on to the next sections of this book, keep your diagnostic test results in mind. Use them to guide your studies, focus on your weak areas, and track your progress. This book is designed to provide you with the knowledge, strategies, and practice you need to succeed, and it all begins with understanding where you stand today.

Bonus Section: Take Your Diagnostic Test Online

As you begin your SAT preparation journey, it's important to start with a solid understanding of your current strengths and areas for improvement. Our comprehensive Diagnostic Test is designed to help you assess your skills across all sections of the SAT, including Math, Evidence-Based Reading and Writing, and the optional Essay.

This diagnostic test will give you a clear picture of where you stand and what you need to focus on as you study. By taking this test before diving into the content, you'll be able to tailor your study plan to your specific needs, ensuring that you spend your time efficiently and effectively.

By starting with the diagnostic test, you'll set yourself up for a more personalized and effective study experience. Good luck, and remember, this is just the first step on your path to SAT success!

Access the Full Diagnostic Test

To make it easier for you to take the Diagnostic Test, we've made it available online. You can take the test at your convenience, track your progress, and receive immediate feedback on your performance.

- **Link to the Diagnostic Test**: https://ueseducation.typeform.com/diagnostic?typeform-source=www.ueseducation.com
- **QR Code for Easy Access**:

Simply scan the QR code or follow the link to access the test. Once you've completed it, you'll receive a detailed report that breaks down your results by section, helping you identify the areas where you need to focus your studies.

Why Take the Diagnostic Test?

- **Identify Strengths and Weaknesses**: Understand where you excel and where you need to improve.
- **Tailor Your Study Plan**: Focus on the areas that will have the biggest impact on your overall score.
- **Track Your Progress**: Use your diagnostic results as a baseline to measure your improvement over time.

Section 1: Math

Math Fundamentals

Overview of Math Sections

The Math section of the SAT is designed to assess your understanding of mathematical concepts, your ability to solve problems, and your capacity to apply these concepts in real-world scenarios. This section is divided into two parts: one that allows the use of a calculator and one that does not. The questions cover a range of topics from algebra, geometry, trigonometry, and data analysis.

Key Concepts and Formulas

In order to excel in the SAT Math section, it's essential to have a solid grasp of key concepts and formulas. Here are some of the most important ones you need to know:

Pythagorean Theorem:

$$c = \sqrt{a^2 + b^2}$$

Quadratic Formula:

$$x = \frac{-b \pm \sqrt{b^2 - 4ac}}{2a}$$

Slope of a Line:

$$m = \frac{y_2 - y_1}{x_2 - x_1}$$

Area of a Circle:

$$A = \pi r^2$$

Circumference of a Circle:

$$C = 2\pi r$$

Volume of a Cylinder:

$$V = \pi r^2 h$$

Heart of Algebra

Algebra is a critical component of the SAT Math section. It includes topics such as linear equations, inequalities, and systems of equations.

Linear Equations and Inequalities

Linear equations are equations of the first degree, meaning they involve only the first power of the variable. A standard form of a linear equation is:

$$ax + by = c$$

Linear inequalities, on the other hand, express a relationship in which one side is not necessarily equal to the other:

$$ax + b \leq c$$

Systems of Equations

A system of equations consists of two or more equations with the same set of variables. The solution to the system is the set of values that satisfy all equations simultaneously. For example:

$$2x + 3y = 6$$
$$x - y = 2$$

To solve this system, you can use substitution, elimination, or graphical methods.

Word Problems

Word problems require translating real-world situations into mathematical expressions. For instance, if a car travels 300 miles at a speed of 60 miles per hour, the time taken can be calculated using:

$$textTime = \frac{\text{Distance}}{\text{Speed}} = \frac{300}{60} = 5 \text{ hours}$$

Practice Problems with Solutions

Here are some practice problems to help reinforce your understanding:

Solve the equation for x:

$$3x - 7 = 11$$

Solution:

$$x = \frac{11 + 7}{3} = 6$$

Solve the system of equations:

$$x + y = 5$$

$$2x - y = 3$$

Solution: Add the equations to eliminate y:

$$3x = 8 \quad \Rightarrow \quad x = \frac{8}{3}$$

Substitute x into the first equation to find y:

$$\frac{8}{3} + y = 5 \quad \Rightarrow \quad y = 5 - \frac{8}{3} = \frac{7}{3}$$

Problem Solving and Data Analysis

Problem Solving and Data Analysis requires a strong understanding of ratios, proportions, percentages, and basic statistics.

Ratios, Proportions, and Percentages

Ratio: A ratio is a comparison of two quantities. For example, the ratio of a to b can be written as:

$$\text{Ratio} = \frac{a}{b}$$

Proportion: A proportion states that two ratios are equal:

$$\frac{a}{b} = \frac{c}{d}$$

Percentage: A percentage is a way of expressing a number as a fraction of 100:

$$\text{Percentage} = \frac{\text{Part}}{\text{Whole}} \times 100\backslash\%$$

Data Interpretation and Graphs

Data interpretation involves analyzing graphs, charts, and tables. You might be asked to extract information from a graph, such as identifying trends or making predictions.

For example, if a bar graph shows the sales of a company over five years, you may be asked to determine the year with the highest sales.

Statistics and Probability

Mean (Average):

$$\text{Mean} = \frac{\sum \text{Values}}{\text{Number of Values}}$$

Median: The median is the middle value when the data set is ordered from least to greatest.

Probability:

$$P(\text{Event}) = \frac{\text{Number of Favorable Outcomes}}{\text{Total Number of Outcomes}}$$

Passport to Advanced Math

Advanced Math topics include complex equations and functions that require deeper understanding.

Complex Equations and Functions

Complex equations might involve multiple steps or the use of several mathematical operations. An example is solving quadratic equations using the quadratic formula mentioned earlier:

$$x = \frac{-b \pm \sqrt{b^2 - 4ac}}{2a}$$

Polynomials and Rational Expressions

Polynomials are expressions that involve sums of powers of xxx with coefficients. A standard polynomial expression looks like:

$$ax^n + bx^{n-1} + \cdots + k$$

where n is a non-negative integer, and $a, b, \ldots, ka, b, \ldots, ka, b, \ldots, k$ are constants.

Exponential and Quadratic Functions

Exponential functions have the form:

$$y = a \cdot b^x$$

where b is a positive constant.

Quadratic functions, as mentioned earlier, follow the form:

$$y = ax^2 + bx + c$$

Additional Topics in Math

Additional topics include geometry, trigonometry, and more advanced mathematical concepts.

Geometry and Trigonometry

In geometry, you'll work with shapes, sizes, and the properties of space. For instance, the area of a triangle is given by:

$$A = \frac{1}{2} \times \text{base} \times \text{height}$$

Trigonometry deals with the relationships between the angles and sides of triangles. For example, the sine of an angle is:

$$\sin(\theta) = \frac{\text{Opposite Side}}{\text{Hypotenuse}}$$

Coordinate Geometry

Coordinate geometry involves plotting points, lines, and curves on a plane. The distance between two points $(x1, y1)(x\_1, y\_1)(x1, y1)$ and $(x2, y2)(x\_2, y\_2)(x2, y2)$ can be found using:

$$d = \sqrt{(x_2 - x_1)^2 + (y_2 - y_1)^2}$$

Core Math Concepts and Practice

The SAT Math section is designed to test a wide range of mathematical concepts, from basic arithmetic to more advanced topics like algebra, geometry, and data analysis. Understanding these core concepts is essential for tackling the variety of questions you'll encounter. In this section, we'll delve into key mathematical topics, explore common types of questions, and provide practice problems with solutions. The goal is to strengthen your understanding of each concept and help you apply these concepts in test scenarios.

Algebra: The Foundation of SAT Math

Algebra forms the basis of many questions on the SAT. Whether it's solving for variables, manipulating equations, or understanding functions, algebraic skills are crucial.

Understanding and Solving Linear Equations

Linear equations are among the most common types of problems on the SAT. A linear equation can be written in the form:

$$ax + by = c$$

To solve for x, isolate the variable by manipulating the equation. For example, if you're given:

$$3x + 4 = 19$$

Subtract 4 from both sides:

$$3x = 15$$

Then, divide both sides by 3:

$$x = 5$$

Systems of Equations

A system of equations consists of two or more equations that share the same variables. The solution is the point(s) where the equations intersect, meaning the values that satisfy all equations simultaneously.

For example:

$$2x + y = 8$$

$$x - y = 1$$

To solve, add the equations to eliminate y:

$$3x = 9 \quad \Rightarrow \quad x = 3$$

Substitute x=3x = 3x=3 into the second equation to find y:

$$3 - y = 1 \quad \Rightarrow \quad y = 2$$

Thus, the solution is (3,2).

Quadratic Equations and Factoring

Quadratic equations often appear in the SAT Math section. They typically have the form:

$$ax^2 + bx + c = 0$$

To solve quadratics, you can factor the equation, complete the square, or use the quadratic formula:

$$x = \frac{-b \pm \sqrt{b^2 - 4ac}}{2a}$$

Example: Solve the quadratic equation:

$$x^2 - 5x + 6 = 0$$

Factor the equation:

$$(x - 2)(x - 3) = 0$$

Thus, the solutions are x=2x = 2x=2 and x=3x = 3x=3.

Advanced Algebra: Functions and Inequalities

Beyond basic algebra, the SAT also tests your understanding of functions and inequalities.

Functions and Their Properties

A function relates an input to a specific output. For example, in the function f(x)=2x+3f(x) = 2x + 3f(x)=2x+3, the input x determines the output f(x). You might be asked to evaluate a function for a given value of x or to solve for x when f(x) is known.

Example: If f(x)=3x−7f(x) = 3x - 7f(x)=3x−7, find f(2):

$$f(2) = 3(2) - 7 = 6 - 7 = -1$$

Solving Inequalities

Inequalities are similar to equations but instead of an equal sign, they use inequality symbols like $<, \leq, >, or\ \geq$.

Example: Solve the inequality:

$$2x - 3 < 7$$

Add 3 to both sides:

$$2x < 10$$

Then, divide both sides by 2:

$$x < 5$$

Geometry and Measurement: Visual and Spatial Reasoning

The SAT includes a variety of geometry questions that test your ability to reason with shapes, angles, areas, volumes, and more.

Triangles and Their Properties

Triangles are a key focus area, especially right triangles. Remember the Pythagorean Theorem for right triangles:

$$c = \sqrt{a^2 + b^2}$$

where ccc is the hypotenuse, and a and b are the other two sides.

Example: In a right triangle, if a=6a = 6a=6 and b=8b = 8b=8, find c:

$$c = \sqrt{6^2 + 8^2} = \sqrt{36 + 64} = \sqrt{100} = 10$$

Circles and Their Equations

The equation of a circle in the coordinate plane with center (h,k)(h, k)(h,k) and radius rrr is:

$$(x - h)^2 + (y - k)^2 = r^2$$

Example: Write the equation of a circle with center (3,−2) and radius 5:

$$(x-3)^2+(y+2)^2=25$$

Volume and Surface Area

Volume and surface area questions often involve cylinders, cones, spheres, and rectangular prisms. Know the formulas:

Cylinder Volume:

$$V=\pi r^2 h$$

Sphere Surface Area:

$$A=4\pi r^2$$

Example: Find the volume of a cylinder with radius 3 and height 7:

$$V=\pi(3)^2(7)=63$$

Data Analysis: Interpreting and Analyzing Information

The SAT tests your ability to work with data, including interpreting graphs, understanding statistical measures, and solving problems based on data sets.

Mean, Median, Mode, and Range

Understanding basic statistics is essential. The mean is the average, the median is the middle value, the mode is the most frequent value, and the range is the difference between the highest and lowest values.

Example: Find the mean, median, mode, and range of the data set 3,7,7,8,10

Mean:

$$\text{Mean} = \frac{3 + 7 + 7 + 8 + 10}{5} = 7$$

Median: 7 (middle value)

Mode: 7 (appears most frequently)

Range:

$$\text{Range} = 10 - 3 = 7$$

Interpreting Graphs and Tables

Graphs, charts, and tables present data visually. You may be asked to interpret information, identify trends, or make predictions based on the data.

Example: A bar graph shows the number of books read by students over five months. If the bars for January, February, and March are 5, 7, and 9 respectively, what is the average number of books read during these months?

$$\text{Average} = \frac{5 + 7 + 9}{3} = 7$$

Probability and Combinatorics

Probability questions on the SAT often involve simple calculations of likelihood, while combinatorics problems deal with counting possibilities.

Basic Probability:

$$P(\text{Event}) = \frac{\text{Number of Favorable Outcomes}}{\text{Total Number of Outcomes}}$$

Example: What is the probability of rolling a 4 on a six-sided die?

$$P(\text{Rolling a 4}) = \frac{1}{6}$$

Combinations:

To calculate the number of ways to choose r objects from n objects without regard to order:

$$\binom{n}{r} = \frac{n!}{r!\,(n-r)!}$$

Example: How many ways can you choose 2 students from a group of 5?

$$\binom{5}{2} = \frac{5!}{2!\,(5-2)!} = \frac{5 \times 4}{2 \times 1} = 10$$

Practice Problems and Solutions

Here are a few practice problems to help solidify these concepts:

1. Linear Equation: Solve for x in the equation:

$$4x - 5 = 3x + 7$$

Solution: Subtract 3x from both sides:

$$x - 5 = 7 \quad \Rightarrow \quad x = 12$$

2. Quadratic Equation: Solve the quadratic equation:

$$x^2 - 4x - 12 = 0$$

Solution: Factor the equation:

$$(x-6)(x+2) = 0 \quad \Rightarrow \quad x = 6 \text{ or } x = -2$$

3. Probability: A jar contains 3 red, 4 blue, and 5 green marbles. What is the probability of randomly selecting a blue marble?

$$P(\text{Blue}) = \frac{4}{3+4+5} = \frac{4}{12} = \frac{1}{3}$$

Mastering the core math concepts covered in the SAT requires both understanding and practice. By familiarizing yourself with these topics and working through practice problems, you'll build the skills needed to tackle the SAT Math section with confidence. Remember to review any areas where you feel less confident and continue practicing until you feel comfortable with each type of question. Good luck with your preparation!

Desmos Calculator: A Powerful Tool for the SAT Math Section

The Desmos calculator is an advanced graphing calculator tool that is available for use during the SAT Math section. Unlike traditional handheld calculators, Desmos offers a user-friendly interface with powerful capabilities that can help you visualize and solve complex problems more effectively. In this section, we'll explore how to use Desmos to your advantage on the SAT, covering essential functions, tips, and strategies for getting the most out of this tool.

Getting Started with Desmos

Before diving into specific tips and tricks, it's important to familiarize yourself with the Desmos interface. The Desmos calculator is available online and is embedded within the SAT's digital testing platform, so you should practice using it in a browser to get comfortable with its features.

1. **Basic Interface Overview**:
    - **Graphing Area**: The large, central area where graphs of equations and data points are displayed.
    - **Expression List**: Located on the left side of the screen, this is where you input equations, inequalities, and other expressions.
    - **Toolbox**: Contains buttons for common mathematical functions, such as square roots, exponents, and trigonometric functions.
2. **Creating and Editing Graphs**:
    - To plot an equation, simply type it into the expression list and hit enter. Desmos will automatically graph the equation in the graphing area.
    - You can adjust the viewing window by clicking and dragging on the graph or by zooming in and out using the mouse wheel or zoom buttons.
3. **Navigating the Graph**:
    - **Pan and Zoom**: Drag the graph to pan around the coordinate plane and use the plus (+) and minus (-) buttons or scroll to zoom in and out.
    - **Resetting the View**: If you lose track of your graph, you can quickly reset the view to fit the entire graph on the screen by clicking the home button.

Key Features and Functions

Desmos is more than just a graphing tool; it's a robust calculator that can handle a wide range of mathematical operations. Here's how to use some of the key features that are particularly useful for the SAT:

1. **Graphing Functions and Equations**:
   - **Linear Equations**: For any linear equation like y=2x+3y = 2x + 3y=2x+3, Desmos will instantly graph the line, showing both the slope and y-intercept.
   - **Quadratic Equations**: Graph quadratics such as y=x2−4x+4y = x^2 - 4x + 4y=x2−4x+4 to find the vertex, roots, and axis of symmetry. You can also use Desmos to visualize the effect of changing coefficients on the shape of the parabola.
2. **Solving Systems of Equations**:
   - Input multiple equations into the expression list, and Desmos will graph each one on the same coordinate plane. The intersection points represent the solutions to the system.
   - For example, entering y=2x+1y = 2x + 1y=2x+1 and y=−x+4y = -x + 4y=−x+4 will show where the two lines intersect, which is the solution to the system.
3. **Inequalities**:
   - Desmos can also handle inequalities. Enter an inequality like y≤2x+3y \leq 2x + 3y≤2x+3, and the calculator will shade the region that satisfies the inequality.
   - You can use this feature to solve systems of inequalities visually, identifying the feasible region where all conditions are met.
4. **Exploring Transformations**:
   - Desmos makes it easy to explore transformations such as translations, reflections, and dilations. By manipulating the coefficients and constants in your equations, you can see how the graph changes in real-time.
   - This is particularly useful for understanding concepts like the vertex form of a quadratic function or the effects of altering the slope and y-intercept in linear functions.
5. **Using the Table Feature**:
   - Desmos allows you to create tables of values, which can be particularly useful for plotting points or analyzing data.
   - To create a table, click the "+" button in the expression list and select "Table." You can enter specific xxx values and Desmos will automatically calculate the corresponding yyy values based on your function.
6. **Working with Trigonometric Functions**:
   - Trigonometric functions like sine, cosine, and tangent can be graphed by entering them directly into Desmos. For example, typing y=sin⁡(x)y = \sin(x)y=sin(x) will graph the sine wave.
   - You can also explore trigonometric identities and transformations by adding phase shifts, amplitude changes, or period adjustments to these functions.
7. **Calculating Slopes and Intercepts**:
   - To quickly find the slope or y-intercept of a linear equation, simply graph the equation and click on the graph. Desmos provides key information about the line, including slope and intercept values.
   - This is especially useful for problems that ask you to interpret linear models or compare the steepness of different lines.
8. **Finding Intersections and Roots**:
   - When dealing with quadratic equations or any function where you need to find roots (zeroes), graph the function and look for the points where it crosses the x-axis.
   - Desmos will highlight these points, making it easy to identify the roots. You can also find the intersection points between different functions or curves.
9. **Exploring Data with Regression**:
   - Desmos includes a regression feature that allows you to fit a line or curve to a set of data points. This is useful for interpreting scatterplots and understanding the relationship between variables.
   - Enter your data points into a table, and then use the regression tools to find the best fit line or curve, along with the equation of the line and the correlation coefficient.

Desmos for Specific SAT Math Problems

Now that you're familiar with the Desmos calculator's capabilities, let's look at how you can apply these features to different types of SAT Math problems:

1. **Linear and Quadratic Word Problems**:
   - Use Desmos to quickly graph equations derived from word problems. This helps you visualize the problem, identify intercepts, and determine the relationship between variables.
   - For example, if a problem involves finding the maximum height of a projectile, graph the quadratic equation representing the height as a function of time. The vertex of the parabola will give you the maximum height.
2. **Geometry and Trigonometry**:
   - For problems involving circles, parabolas, or other shapes, graph the equations and use Desmos to find key points like intersections, tangents, and distances.
   - In trigonometry problems, graph sine and cosine functions to analyze their properties or solve equations involving angles and periods.
3. **Data Analysis**:
   - Use Desmos to create scatterplots from data sets provided in SAT Math questions. You can then apply regression analysis to find trends and make predictions based on the data.
   - This is particularly useful for questions that ask you to interpret data or predict future values based on a given model.
4. **Systems of Equations and Inequalities**:
   - Graphing systems of equations and inequalities on Desmos allows you to easily find the solution set, which is represented by the intersection of the graphs or the overlapping shaded regions.
5. **Function Analysis**:
   - When dealing with functions, Desmos can help you quickly identify critical points, such as maxima, minima, and points of inflection, by graphing the function and analyzing its behavior.

Tips for Using Desmos Efficiently

To make the most of Desmos during the SAT, keep these tips in mind:

1. **Practice Before the Test**: Spend time using Desmos before test day so you're comfortable with its features. Familiarize yourself with the shortcuts and practice solving a variety of problems.
2. **Don't Overuse**: While Desmos is a powerful tool, don't rely on it for every question. Some problems can be solved more quickly using mental math or simple calculations.
3. **Use Desmos to Check Work**: After solving a problem by hand, use Desmos to check your work. For example, if you've solved a system of equations, graph both equations to ensure they intersect at the point you found.
4. **Stay Organized**: Keep your Desmos workspace organized by labeling equations and using color coding to distinguish between different graphs. This helps avoid confusion during the test.
5. **Understand What You're Graphing**: Always have a clear understanding of the mathematical concept before graphing. Desmos can help you visualize and confirm your solution, but it's important to know why the graph behaves as it does.

6. **Time Management**: Desmos can save time on complex problems, but it's important to balance its use with time management. Don't spend too much time tweaking graphs; get the information you need and move on.

Bonus Section: Step-by-Step Tutorial for Desmos Calculator

The Desmos calculator is a powerful tool that can greatly enhance your performance on the SAT Math section. Whether you need to graph equations, solve systems of equations, or explore functions and their transformations, Desmos can help you visualize and solve problems quickly and efficiently. To ensure you're getting the most out of this tool, we've created a step-by-step tutorial that walks you through the essential features and functions of Desmos.

In this tutorial, you'll learn how to:

- Graph and analyze linear, quadratic, and trigonometric functions.
- Solve systems of equations and inequalities.
- Use tables to analyze data and find lines of best fit.
- Apply transformations to functions and understand their effects.

Access the Video Tutorial

To help you master Desmos, we've created a detailed video tutorial that takes you through each step in an easy-to-follow format. You can watch the tutorial by scanning the QR code below or by following the link provided. This video will guide you through real SAT-style problems, showing you exactly how to use Desmos to find solutions and gain a deeper understanding of the material.

By practicing with this tutorial, you'll become more confident in your ability to use Desmos on test day, giving you an edge in tackling even the most challenging SAT Math questions. Don't forget to bookmark the video so you can refer back to it as you continue your studies!

The Desmos calculator is an invaluable tool for tackling the SAT Math section, offering powerful features that can help you solve and visualize problems more effectively. By practicing with Desmos, understanding its capabilities, and knowing when and how to use it during the test, you can enhance your problem-solving skills and boost your SAT Math score. Remember, the key to success with Desmos is practice and familiarity, so make sure to integrate it into your study routine.

Math Practice Test: No Calculator Section

Instructions:

- You will have 25 minutes to complete this section.
- Solve each problem and select the best answer from the choices provided.
- For grid-in questions, write your answer in the boxes provided.

Questions:

1. What is the value of $3x + 4$ when $x = 2$?

- A) 6
- B) 8
- C) 10
- D) 14

2. If $5y - 3 = 2y + 12$, what is the value of y?

- A) 3
- B) 4
- C) 5
- D) 6

3. What is the slope of the line passing through the points (1,2) and (4,8)?

- A) $\frac{2}{3}$
- B) 2
- C) $\frac{3}{2}$
- D) $\frac{4}{7}$

4. Simplify the expression: $2(3x - 5) + 4x$.

- A) $10x - 10$
- B) $6x - 5$
- C) $6x + 10$
- D) $10x - 5$

5. If $2a + 3b = 10$ and $a = 2$, what is the value of b?

- A) 1
- B) 2
- C) 3

- D) 4

6. Solve for x in the equation $4x - 7 = 9$.

- A) $x = 2$
- B) $x = 3$
- C) $x = 4$
- D) $x = 5$

7. Which of the following is equivalent to $5(x + 2)$?

- A) $5x + 2$
- B) $5x + 10$
- C) $5x + 7$
- D) $5x + 12$

8. If the area of a square is 49 square units, what is the length of one side of the square?

- A) 5
- B) 6
- C) 7
- D) 8

9. What is the value of xxx in the equation $3x + 2 = 11$?

- A) 1
- B) 2
- C) 3
- D) 4

10. Simplify the expression: $12x - 4(2x + 1)$.

- A) $4x - 4$
- B) $8x - 4$
- C) $4x + 4$
- D) $8x + 4$

11. If $y = 2x + 3$, what is the value of y when $x = -1$?

- A) 1
- B) 3
- C) 5
- D) 7

12. Which of the following represents the equation of a line with a slope of $-\frac{1}{2}$ and a y-intercept of 4?

- A) $y = -\frac{1}{2}x + 4$

- B) $y = \frac{1}{2}x + 4$
- C) $y = -2x + 4$
- D) $y = -\frac{1}{2}x - 4$

13. The sum of three consecutive integers is 51. What is the smallest of these integers?

- A) 16
- B) 17
- C) 18
- D) 19

14. If $4x - 2y = 8$, what is the value of y when $x = 3$?

- A) 1
- B) 2
- C) 3
- D) 4

15. What is the solution to the inequality $2x + 3 < 7$?

- A) $x < 2$
- B) $x > 2$
- C) $x < 1$
- D) $x > 1$

16. If the perimeter of a rectangle is 30 units and the length is 10 units, what is the width?

- A) 5 units
- B) 7 units
- C) 8 units
- D) 10 units

17. What is the value of $\frac{3x}{2} + 4 = 10$?

- A) 2
- B) 3
- C) 4
- D) 6

18. If $p(x) = x^2 + 3x - 4$, what is the value of $p(2)$?

- A) 2
- B) 4
- C) 6
- D) 8

19. Solve the system of equations:

$$2x + y = 7$$

$$x - y = -1$$

- A) x=2x = 2x=2, y=3y = 3y=3
- B) x=3x = 3x=3, y=1y = 1y=1
- C) x=1x = 1x=1, y=2y = 2y=2
- D) x=4x = 4x=4, y=3y = 3y=3

20. Simplify the expression $(3x^2y)(2xy^3)$

- A) $6x^3y^4$
- B) $5x^2y^3$
- C) $6x^2y^3$
- D) $5x^3y^4$

Math Practice Test: Calculator Section

Instructions:

- You will have 55 minutes to complete this section.
- Solve each problem and select the best answer from the choices provided.
- For grid-in questions, write your answer in the boxes provided.

Questions:

1. Solve for x in the equation $5x^2 - 3x + 2 = 0$.

- A) $x = 1$
- B) $x = \frac{1}{2}$
- C) $x = -\frac{2}{5}$
- D) $None\ of\ the\ above$

2. A car travels 150 miles using 5 gallons of gasoline. What is the car's fuel efficiency in miles per gallon?

- A) 20 mpg
- B) 25 mpg
- C) 30 mpg
- D) 35 mpg

3. If $2x + 3y = 12$ and $4x - y = 5$, what is the value of x?

- A) 1
- B) 2
- C) 3
- D) 4

4. The function $f(x) = 3x + 7$ is graphed in the xy-plane. What is the slope of the line?

- A) 3
- B) 7
- C) $\frac{7}{3}$
- D) $\frac{3}{7}$

5. A rectangular garden has a length of 12 meters and a width of 9 meters. What is the area of the garden?

- A) $96m^2$

- B) $108m^2$
- C) $144m^2$
- D) $180m^2$

6. What is the value of $\log_2(32)$?

- A) 4
- B) 5
- C) 6
- D) 7

7. If the volume of a cube is 64 cubic centimeters, what is the length of one side of the cube?

- A) 2 cm
- B) 4 cm
- C) 8 cm
- D) 16 cm

8. The equation $y = x^2 - 4x + 6$ represents a parabola. What is the y-coordinate of the vertex of the parabola?

- A) -2
- B) 0
- C) 2
- D) 4

9. A right triangle has legs of length 6 and 8. What is the length of the hypotenuse?

- A) 8
- B) 10
- C) 12
- D) 14

10. Solve for x in the inequality $4x - 7 > 5$.

- A) $x > 1$
- B) $x > 2$
- C) $x > 3$
- D) $x > 4$

11. The circle $x^2 + y^2 = 49$ is graphed in the xy-plane. What is the radius of the circle?

- A) 5
- B) 6
- C) 7
- D) 8

12. A tank is filling with water at a rate of 3 liters per minute. If the tank is currently holding 20 liters of water, how much water will it hold after 10 more minutes?

- A) 50 liters
- B) 45 liters
- C) 55 liters
- D) 60 liters

13. The sum of the angles in a triangle is 180 degrees. If one angle is 40 degrees and another is 60 degrees, what is the measure of the third angle?

- A) 70 degrees
- B) 80 degrees
- C) 90 degrees
- D) 100 degrees

14. If the length of a rectangle is increased by 20% and the width is decreased by 10%, what is the percentage change in the area of the rectangle?

- A) 8% increase
- B) 8% decrease
- C) 10% increase
- D) 10% decrease

15. The distance between two points (x_1, y_1) and (x_2, y_2) in the xy-plane is given by the distance formula. What is the distance between the points $(3, 4)$ and $(7, 1)$?

- A) 4
- B) 5
- C) 6
- D) 7

16. A company's profit is modeled by the function $P(x) = -5x^2 + 150x - 200$, where x is the number of units sold. What number of units sold maximizes the company's profit?

- A) 10
- B) 15
- C) 20
- D) 30

17. What is the value of $\sin 30°$?

- A) $\frac{1}{2}$
- B) $\frac{\sqrt{2}}{2}$
- C) $\frac{\sqrt{3}}{2}$
- D) 1

18. If $2^x = 8$, what is the value of x?

- A) 2
- B) 3
- C) 4
- D) 5

19. A sequence is defined by the formula $a_n = 3n - 2$. What is the 5th term in the sequence?

- A) 10
- B) 11
- C) 12
- D) 13

20. The product of two consecutive even integers is 48. What are the integers?

- A) 4 and 6
- B) 6 and 8
- C) 8 and 10
- D) 10 and 12

21. If $f(x) = 2x^2 - 3x + 5$, what is the value of $f(3)$?

- A) 8
- B) 11
- C) 14
- D) 17

22. What is the sum of the solutions to the equation $x^2 - 5x + 6 = 0$?

- A) 2
- B) 3
- C) 5
- D) 6

23. The radius of a circle is doubled. What is the effect on the area of the circle?

- A) The area remains the same.
- B) The area is doubled.
- C) The area is quadrupled.
- D) The area is halved.

24. A cylinder has a height of 10 cm and a radius of 3 cm. What is the volume of the cylinder? (Use $\pi \approx 3.14$)

- A) $94.2cm^3$
- B) $282.6cm^3$
- C) $565.2cm^3$
- D) $942.0cm^3$

25. Solve for x: $3^{2x} = 81$.

- A) x=2
- B) x=3
- C) x=4
- D) x=5

26. If $x + 2 = 3y - 4$, what is y in terms of x?

- A) $y = \frac{x+6}{3}$
- B) $y = \frac{x-2}{3}$
- C) $y = \frac{x+4}{3}$
- D) $y = \frac{x-6}{3}$

27. A box contains 3 red, 4 blue, and 5 green balls. If a ball is selected at random, what is the probability that it is either red or green?

- A) $\frac{3}{12}$
- B) $\frac{5}{12}$
- C) $\frac{8}{12}$
- D) $\frac{9}{12}$

28. A triangle has sides of length 5, 12, and 13. What is the area of the triangle?

- A) 30
- B) 31
- C) 32
- D) 33

29. If a and b are inversely proportional, and $a = 4$ when $b = 6$, what is the value of a when $b = 8$?

- A) 2
- B) 3
- C) 4
- D) 6

30. The function $h(t) = -16t^2 + 32t + 50$ represents the height of an object after t seconds. What is the maximum height reached by the object?

- A) 50 feet
- B) 66 feet
- C) 74 feet
- D) 82 feet

31. If x and y are integers such that $x + y = 7$ and $x - y = 3$, what is the value of $x \times y$?

- A) 10
- B) 11
- C) 12
- D) 13

32. A circle has a circumference of 20 cm. What is the diameter of the circle? (Use $\pi \approx 3.14$)

- A) 5.7 cm
- B) 6.4 cm
- C) 7.2 cm
- D) 8.1 cm

33. The function $g(x)$ is defined as $g(x) = \frac{1}{x}$. What is the value of $g(-4)$?

- A) -4
- B) $-\frac{1}{4}$
- C) $\frac{1}{4}$
- D) 4

34. The inequality $3x - 2 \leq 7$ is equivalent to which of the following?

- A) $x \leq 1$
- B) $x \leq 3$
- C) $x \leq 4$
- D) $x \leq 5$

35. In a parallelogram, opposite sides are equal in length. If one side is 8 cm and the adjacent side is 12 cm, what is the perimeter of the parallelogram?

- A) 20 cm
- B) 32 cm
- C) 40 cm
- D) 48 cm

36. A rectangle's length is 4 times its width. If the perimeter of the rectangle is 90 units, what is the area of the rectangle?

- A) 80 square units
- B) 100 square units
- C) 180 square units
- D) 225 square units

37. The population of a town is increasing by 5% each year. If the current population is 20,000, what will the population be in 3 years?

- A) 21,000
- B) 22,500
- C) 23,150

- D) 23,315

38. What is the value of $\sqrt{144} - \sqrt{81}$?

- A) 3
- B) 5
- C) 7
- D) 9

SAT Math Practice Test Answer Key: No Calculator and Calculator Sections

No Calculator Section

Q1	Q2	Q3	Q4	Q5	Q6	Q7	Q8	Q9	Q10
D	C	B	A	C	D	B	C	C	A

Q11	Q12	Q13	Q14	Q15	Q16	Q17	Q18	Q19	Q20
C	A	B	B	A	A	C	B	B	A

Calculator Section

Q1	Q2	Q3	Q4	Q5	Q6	Q7	Q8	Q9	Q10
D	C	B	A	C	D	B	C	B	A

Q11	Q12	Q13	Q14	Q15	Q16	Q17	Q18	Q19	Q20
C	A	B	C	B	C	A	B	C	A

Q21	Q22	Q23	Q24	Q25	Q26	Q27	Q28	Q29	Q30
D	C	C	B	A	A	C	A	B	C

Q31	Q32	Q33	Q34	Q35	Q36	Q37	Q38
C	A	B	C	B	C	D	C

Math Strategies and Tips

The SAT Math section can be a challenging part of the test, but with the right strategies, you can maximize your score. This section will cover a variety of tips and techniques that will help you tackle the Math section more effectively. From time management to specific problem-solving tactics, these strategies are designed to give you the confidence and skills needed to perform your best.

Time Management

One of the most critical aspects of the SAT Math section is managing your time efficiently. The test is divided into two parts: a 25-minute section without a calculator and a 55-minute section with a calculator. Here's how you can manage your time effectively:

1. **Prioritize Questions**: Not all questions are created equal. Some will be easier for you than others. Skim through the questions at the beginning and identify the ones you find most straightforward. Start with those to build confidence and secure easy points.
2. **Pacing**: Aim to spend no more than a minute or two on each question during your first pass. If a problem seems too difficult or time-consuming, skip it and return to it later if you have time.
3. **Use the Process of Elimination**: When you're unsure of an answer, eliminate the choices that are clearly wrong. This increases your chances of guessing correctly and saves time by narrowing down your options.
4. **Watch the Clock**: Regularly check the time so you're aware of how much you have left. However, don't become overly focused on the clock. Balance is key—be mindful of time but not distracted by it.
5. **Save Time for Review**: Ideally, try to finish the section with a few minutes to spare. Use this time to review your answers, especially the questions you were unsure about. Double-check your calculations and ensure you haven't misread any questions.

Calculator Tips (Including Desmos)

Using a calculator effectively can significantly improve your performance in the SAT Math section. While you're allowed to use a calculator for some questions, knowing when and how to use it is crucial.

1. **Know Your Calculator**: Whether you're using a graphing calculator, a scientific calculator, or an online tool like Desmos, familiarity is key. Spend time practicing with your calculator before the test so you're comfortable with its functions.
2. **When to Use the Calculator**: Not every problem requires a calculator. For instance, basic arithmetic, simple fractions, and certain algebraic manipulations can be done faster by hand. Use your calculator for complex calculations, graphing functions, or checking your work.
3. **Avoid Over-Reliance**: It's easy to become dependent on the calculator for every question, but this can slow you down. Understand the underlying math concepts so you can quickly determine when calculator use is necessary.
4. **Graphing Tools**: If your calculator has graphing capabilities, use it to visualize equations and check solutions for functions, especially when dealing with quadratics, systems of equations, or inequalities.
5. **Desmos Tips**: If you're using Desmos, take advantage of its graphing features, table functions, and the ability to quickly plot multiple functions. However, ensure that you are comfortable using it efficiently under timed conditions.

Problem-Solving Techniques

The SAT Math section tests not only your knowledge but also your problem-solving abilities. Here are some techniques that can help you solve problems more effectively:

1. **Understand the Question**: Before jumping into calculations, take a moment to fully understand what the question is asking. Identify the relevant information and what you need to find. Misinterpreting the question is a common mistake that can lead to incorrect answers.
2. **Break Down the Problem**: For complex problems, break them down into smaller, more manageable steps. Solve each part one step at a time, which can help prevent errors and make the problem seem less daunting.
3. **Draw Diagrams**: For geometry and word problems, drawing a diagram can provide clarity and make it easier to visualize the problem. Label all known values and mark what you need to find.
4. **Check Units and Conversions**: Pay attention to the units provided in the problem and ensure your answer is in the correct units. Convert measurements if necessary before performing calculations.
5. **Estimate**: If the problem involves large numbers or complicated calculations, try estimating the answer before you begin. This can help you quickly identify if your final answer is reasonable or if you've made a mistake.
6. **Plug In Numbers**: If you're stuck on an algebraic problem, try plugging in numbers for the variables. This can help you understand the problem better and sometimes even lead you directly to the solution.
7. **Work Backwards**: For some multiple-choice questions, it can be helpful to start with the answer choices and work backwards to see which one fits the problem. This technique can be especially useful for word problems or when the algebra seems too complicated.

Handling Difficult Questions

Not every question on the SAT Math section will be easy, and some may be designed to challenge you. Here's how to handle the more difficult questions:

1. **Stay Calm**: It's easy to panic when you encounter a tough problem, but staying calm is crucial. Take a deep breath and approach the problem methodically.
2. **Use Logical Reasoning**: Even if you don't immediately know how to solve a problem, use logical reasoning to eliminate incorrect answers. Sometimes, understanding what a wrong answer would look like can help guide you to the correct one.
3. **Don't Dwell on One Problem**: If you're stuck on a problem and time is ticking, move on. It's better to answer easier questions first and come back to the difficult ones if you have time remaining.
4. **Look for Patterns**: Some problems may involve patterns or relationships that aren't immediately obvious. For instance, in sequences or function problems, identifying a pattern can lead you to the correct answer.
5. **Guessing Strategies**: If you're truly stumped, it's better to guess than to leave a question blank. The SAT does not penalize for wrong answers, so use the process of elimination to make an educated guess.

Common Mistakes and How to Avoid Them

Mistakes are inevitable during the test, but being aware of common pitfalls can help you avoid them.

1. **Misreading the Question**: Carefully read each question to ensure you understand what's being asked. Look out for words like "not" or "except," which can change the meaning of the question entirely.
2. **Skipping Steps**: In a rush to finish, you might be tempted to skip steps in your calculations. However, this increases the likelihood of errors. Write out your work, especially for complex problems, to keep track of your reasoning.
3. **Neglecting to Review**: If time permits, always review your answers. Double-check your calculations, and ensure that your answer makes sense in the context of the question.
4. **Overcomplicating Simple Problems**: Sometimes the SAT includes questions that seem more complicated than they actually are. Avoid overthinking by focusing on the basic math concepts that apply. If a problem seems too difficult, it might be because you're looking for a complicated solution when a simpler one exists.
5. **Using the Wrong Formula**: Memorize key formulas and know when to apply them. Double-check that you're using the correct formula for the problem at hand, especially for geometry questions.

Mental Preparation and Focus

Your mental state during the test can significantly impact your performance. Here are some tips to help you stay focused and sharp:

1. **Get Plenty of Rest**: A good night's sleep before the test is crucial. Fatigue can impair your ability to think clearly and solve problems efficiently.
2. **Eat a Healthy Breakfast**: Eating a nutritious breakfast can help you maintain your energy levels throughout the test. Avoid heavy or sugary foods that might make you feel sluggish.

3. **Stay Hydrated**: Dehydration can lead to headaches and difficulty concentrating. Drink water before the test and consider bringing a bottle with you if allowed.
4. **Positive Mindset**: Enter the test with a positive attitude. Remind yourself that you've prepared well and are capable of doing your best.
5. **Mindfulness and Relaxation**: If you feel stressed during the test, take a few seconds to close your eyes, take deep breaths, and calm your mind. This can help you regain focus and reduce anxiety.

Practice and Preparation

Finally, the best way to improve your SAT Math score is through consistent practice and preparation.

1. **Practice Regularly**: Set aside time each day to practice math problems. Consistency is key to reinforcing the concepts you've learned.
2. **Take Full-Length Practice Tests**: Simulate test day conditions by taking full-length practice tests. This will help you build endurance and get used to the timing of the exam.
3. **Review Mistakes**: After each practice session, review the problems you got wrong. Understand why you made mistakes and learn from them.
4. **Use Official SAT Practice Materials**: Official SAT practice tests and materials are designed to closely resemble the actual exam. Use these resources to familiarize yourself with the test format and question types.
5. **Seek Help When Needed**: If there are concepts or problems you're struggling with, don't hesitate to seek help. This could be from a teacher, tutor, or online resources.
6. **Stay Motivated**: Preparing for the SAT can be a long and challenging process, but staying motivated is key. Set small goals for each study session and reward yourself when you achieve them.

Section 2: Evidence-Based Reading and Writing

Introduction to Evidence-Based Reading and Writing

The Evidence-Based Reading and Writing section of the SAT is designed to measure your ability to analyze and understand written texts, as well as your ability to effectively revise and edit written material. This section is a critical component of the SAT, accounting for half of your total score. It tests not only your reading comprehension skills but also your ability to identify and correct errors in grammar, usage, and sentence structure. In this section of the book, you will learn strategies for tackling the Reading Test, which includes passages from a variety of subjects such as literature, history, social studies, and science. You will also dive into the Writing and Language Test, where you will be asked to improve the clarity, coherence, and effectiveness of passages, focusing on grammar and usage.

Why is the Evidence-Based Reading and Writing Section Important?

The skills tested in this section are essential for success in college and beyond. Whether you are analyzing complex texts in a humanities course or crafting well-structured essays in a social sciences class, the ability to read critically and write effectively is crucial. This section of the SAT is designed to reflect the kind of reading and writing you will do in college, making it an excellent predictor of your readiness for higher education.
Moreover, the Evidence-Based Reading and Writing score is a key factor that colleges consider during the admissions process. A strong performance in this section can significantly enhance your application, demonstrating your proficiency in reading, writing, and critical thinking.

What to Expect in This Section of the Book

In the pages that follow, you will find comprehensive strategies, practice exercises, and tips designed to help you excel in the Evidence-Based Reading and Writing section of the SAT. This section is divided into two main parts: the Reading Test and the Writing and Language Test.

- **Reading Test**: You will learn how to approach different types of passages, identify main ideas and themes, analyze arguments, and interpret evidence. The strategies provided will help you manage your time effectively and improve your accuracy on this test.
- **Writing and Language Test**: This part of the book will guide you through the most common grammar and usage rules tested on the SAT. You will also learn how to improve the organization and clarity of passages, making sure you understand how to identify and correct errors.

How to Use This Section

To get the most out of this section, start by familiarizing yourself with the types of questions you will encounter on the SAT. As you work through the material, take your time with each strategy and practice exercise. Pay attention to the explanations provided for each answer, as these will help you understand why certain choices are correct or incorrect.

Make sure to complete the practice passages and review the answer explanations thoroughly. This will help you identify areas where you need further improvement and give you the confidence to approach similar questions on the actual test.

Finally, take advantage of the full-length practice tests at the end of this section. These tests are designed to simulate the actual SAT experience, helping you build stamina and refine your test-taking strategies.

By the time you complete this section of the book, you will have a solid understanding of the Evidence-Based Reading and Writing portion of the SAT, as well as the skills and strategies needed to achieve a high score.

Reading Test Overview

The Reading Test is a critical component of the SAT that evaluates your ability to understand, interpret, and analyze written passages. This section is designed to measure your reading comprehension skills across a variety of subjects, including literature, history, social studies, and science. The passages you will encounter are carefully selected to reflect the kinds of reading you will do in college, making this section a strong predictor of your academic readiness.

Structure and Timing

The Reading Test consists of 52 multiple-choice questions, and you will have 65 minutes to complete this section. The test is divided into five passages, each followed by a set of questions that test your ability to comprehend and analyze the material.

- **Number of Passages:** 5
- **Number of Questions:** 52
- **Time Allotted:** 65 minutes

Each passage or pair of passages is typically around 500 to 750 words long, and the questions following each passage focus on various aspects of reading comprehension, including the identification of main ideas, the interpretation of words in context, and the analysis of arguments and evidence.

Types of Passages

The Reading Test includes a variety of passage types, each representing a different subject area. Understanding the nature of each passage type will help you approach the questions more effectively.

1. **Literature:**
    - These passages are typically excerpts from novels, short stories, or plays. They often explore themes related to human experience, relationships, and emotions. The questions in this category will test your ability to interpret characters' motivations, analyze narrative techniques, and understand themes and literary devices.
2. **History/Social Studies:**
    - Passages in this category are drawn from historical documents, speeches, or writings on social sciences. They may discuss political, social, or economic issues. Questions will often require you

to analyze the author's argument, understand the historical context, and evaluate the use of evidence.

3. **Science:**
   - Science passages are based on topics in biology, chemistry, physics, or Earth sciences. These passages focus on explaining scientific concepts, theories, or experiments. The questions will test your ability to comprehend complex scientific information, interpret data, and understand the implications of scientific findings.
4. **Paired Passages:**
   - Occasionally, the Reading Test includes paired passages, where two related texts are presented together. These passages might offer different perspectives on a single topic or present contrasting arguments. The questions will ask you to compare and contrast the two passages, analyze their arguments, and evaluate the evidence presented.
5. **Informational Graphics:**
   - Some passages may be accompanied by informational graphics such as charts, graphs, or tables. These graphics are used to present data or illustrate concepts mentioned in the text. You will be required to interpret the information in the graphics and understand how it relates to the passage.

Question Types

The questions on the Reading Test are designed to assess various aspects of your reading comprehension skills. Understanding the types of questions you will encounter can help you approach them strategically.

1. **Detail Questions:**
   - These questions ask you to locate and understand specific information in the passage. You may be asked to identify facts, figures, or details mentioned by the author.
2. **Inference Questions:**
   - Inference questions require you to read between the lines and make logical deductions based on the information presented in the passage. These questions often begin with phrases like "It can be inferred that..." or "The passage suggests that..."
3. **Vocabulary in Context:**
   - Vocabulary questions ask you to determine the meaning of a word or phrase as it is used in the passage. These questions test your ability to understand the word's meaning based on the surrounding context.
4. **Function Questions:**
   - Function questions ask you to determine the purpose of a specific sentence, paragraph, or section of the passage. You may be asked how a particular detail supports the main idea or how a certain paragraph contributes to the overall structure of the text.
5. **Author's Purpose and Point of View:**
   - These questions focus on understanding the author's intent and perspective. You may be asked to identify the author's purpose in writing the passage, their tone, or their attitude towards the subject matter.
6. **Evidence-Based Questions:**
   - Evidence-based questions require you to identify the portion of the text that best supports your answer to a previous question. These questions test your ability to back up your interpretations with direct evidence from the passage.

Strategies for Success

To succeed in the Reading Test, it is crucial to develop strong reading strategies and time management skills. Here are some tips to help you navigate this section effectively:

1. **Active Reading:**
   - As you read each passage, engage with the text by underlining key points, circling unfamiliar words, and jotting down brief notes in the margins. This will help you stay focused and retain important information.
2. **Understand the Main Idea:**
   - Before diving into the questions, ensure you understand the main idea of the passage. This will provide you with a framework to answer detail and inference questions more accurately.
3. **Manage Your Time:**
   - Allocate your time wisely, aiming to spend about 12-13 minutes on each passage, including answering the questions. If you find yourself stuck on a difficult question, move on and come back to it later if time permits.
4. **Answer Based on the Passage:**
   - Always base your answers on the information provided in the passage, even if it contradicts what you know or believe. The SAT Reading Test assesses your ability to interpret the text as it is presented.
5. **Eliminate Wrong Answers:**
   - Use the process of elimination to narrow down your choices. Discard any answers that are clearly incorrect or irrelevant to the question.
6. **Practice Regularly:**
   - Familiarize yourself with the types of passages and questions you will encounter by practicing regularly. The more you practice, the more comfortable you will become with the format and timing of the test.

By understanding the structure and demands of the Reading Test, and by applying these strategies, you can approach this section of the SAT with confidence. In the following chapters, we will explore specific reading comprehension techniques, provide practice passages, and offer detailed explanations to help you master this part of the exam.

Reading Comprehension Strategies

Mastering the Reading Test on the SAT requires more than just the ability to read quickly. It demands a deep understanding of how to approach and analyze different types of passages, how to identify the main ideas, and how to effectively answer the variety of questions that follow each passage. In this section, we'll delve into specific strategies that will help you improve your reading comprehension skills and boost your confidence on test day.

Identifying Main Ideas and Themes

One of the most critical skills on the Reading Test is the ability to quickly and accurately identify the main idea of a passage. The main idea is the central point or argument that the author is trying to convey. It's the thread that ties together all the details and examples in the passage.

Strategies for Identifying Main Ideas:

- **Look for the Thesis Statement:** The main idea is often stated directly in the thesis statement, usually found in the introduction or conclusion of the passage.
- **Pay Attention to Topic Sentences:** Topic sentences at the beginning of paragraphs often highlight key points that contribute to the overall main idea.
- **Summarize Each Paragraph:** After reading each paragraph, briefly summarize it in your mind. This helps you piece together how each part of the passage contributes to the main idea.
- **Ask Yourself:** "What is the author trying to say?" This question can help you focus on the core message of the passage.

Practice Example:

Consider a passage discussing the environmental impact of plastic waste. The main idea might be that plastic pollution poses a severe threat to marine life and ecosystems. The passage might use various examples and data points to support this argument, but the central message remains focused on the dangers of plastic waste.

Analyzing Arguments and Evidence

Many passages on the SAT Reading Test are argumentative, meaning the author is trying to persuade you of a particular point of view. Understanding how to analyze these arguments is crucial for answering questions about the author's reasoning and use of evidence.

Key Elements of an Argument:

- **Claims:** The main points or assertions made by the author.
- **Evidence:** Facts, statistics, examples, and quotes used to support the claims.
- **Reasoning:** The logical connections the author makes between the evidence and the claims.

Strategies for Analyzing Arguments:

- **Identify the Author's Claims:** What is the author trying to convince you of? Look for statements that express the author's main points.
- **Evaluate the Evidence:** Is the evidence relevant and sufficient to support the claims? Consider the quality and quantity of the evidence provided.
- **Examine the Reasoning:** How does the author link the evidence to the claims? Look for any logical fallacies or weaknesses in the argument.

Practice Example:

In a passage arguing for the benefits of renewable energy, the author might claim that wind power is a sustainable alternative to fossil fuels. The evidence could include data on wind energy production, examples of successful wind farms, and expert opinions. To analyze the argument, you would assess whether the evidence effectively supports the claim and whether the author's reasoning is logical and convincing.

Understanding Relationships Between Ideas

The Reading Test often includes questions that ask you to understand the relationships between different ideas within a passage. These relationships might be between sentences, paragraphs, or even between two passages in paired sets.

Common Relationships:

- **Cause and Effect:** One event causes another.
- **Comparison and Contrast:** Two or more ideas are compared or contrasted.

- **Problem and Solution:** A problem is presented, followed by one or more solutions.
- **Sequence:** Events or ideas are presented in a specific order.

Strategies for Identifying Relationships:

- **Look for Transition Words:** Words like "because," "therefore," "however," and "similarly" often signal relationships between ideas.
- **Consider the Structure:** The organization of the passage can provide clues about how ideas are related. For example, a passage that starts with a problem and ends with a solution is likely using a problem-solution structure.
- **Make Connections:** As you read, try to connect the dots between different ideas in the passage. This can help you understand the overall structure and flow of the argument.

Practice Example:

In a passage discussing the effects of climate change, the author might first describe the problem (rising global temperatures) and then present the consequences (melting polar ice, rising sea levels). Recognizing this cause-and-effect relationship is crucial for answering related questions correctly.

Practice Passages with Guided Explanations

One of the most effective ways to prepare for the Reading Test is to practice with actual passages similar to those you will encounter on the SAT. Below are some practice passages, followed by detailed explanations for each question.

Practice Passage 1:

Read the following passage and answer the questions that follow.

"The debate over the use of artificial intelligence (AI) in various industries has intensified in recent years. Proponents argue that AI can significantly increase efficiency and productivity, especially in sectors like manufacturing and healthcare. Critics, however, warn of potential job losses and ethical concerns related to data privacy and decision-making algorithms. As AI technology continues to advance, it is essential to address these challenges while embracing the potential benefits."

Question 1: What is the main idea of the passage?

- A) The use of AI is primarily beneficial for increasing efficiency in industries.
- B) The debate over AI focuses on its potential benefits and challenges.
- C) AI technology is advancing too quickly for society to manage.
- D) Ethical concerns about AI outweigh its benefits.

Explanation: The correct answer is B. The passage discusses both the benefits and challenges of AI, indicating that the main idea is the ongoing debate over its use.

Question 2: Which of the following is an example of a claim made in the passage?

- A) AI will lead to significant job losses.
- B) AI technology is advancing rapidly.
- C) The use of AI can increase efficiency in manufacturing and healthcare.
- D) Data privacy concerns are unfounded.

Explanation: The correct answer is C. The passage claims that AI can increase efficiency in certain industries, which is an assertion made by the author.

Practice Passage 2:

"In the late 19th century, the invention of the telephone revolutionized communication. Prior to its invention, long-distance communication was limited to written letters and telegraphs, which were slow and cumbersome. The telephone allowed people to speak directly with one another, regardless of distance, transforming both personal and business interactions. As the technology spread, it became clear that the telephone was not just a luxury but a necessity in the modern world."

Question 1: What relationship does the author describe between the telephone and previous forms of communication?

- A) The telephone replaced older forms of communication entirely.
- B) The telephone improved upon the limitations of earlier communication methods.
- C) The telephone was less efficient than the telegraph.
- D) The telephone and earlier communication methods were used simultaneously without conflict.

Explanation: The correct answer is B. The passage highlights how the telephone improved communication by allowing direct conversation, which was not possible with letters and telegraphs.

By applying these reading comprehension strategies, you can improve your ability to identify main ideas, analyze arguments, and understand the relationships between ideas in the passages you encounter on the SAT. Regular practice with these strategies will help you become more confident and efficient in answering the variety of questions on the Reading Test. In the next section, we will explore the Writing and Language Test, where you will learn how to improve and correct written passages.

Grammar and Usage Rules

The Writing and Language Test on the SAT heavily focuses on your understanding of grammar and usage. Mastery of these rules is crucial for improving the clarity, precision, and effectiveness of your writing. In this section, we will explore the most commonly tested grammar and usage rules, along with strategies for identifying and correcting errors.

Subject-Verb Agreement

Subject-verb agreement is one of the most fundamental aspects of grammar. It requires that the subject and verb in a sentence agree in number, meaning that singular subjects take singular verbs, and plural subjects take plural verbs.

Basic Rules:

- **Singular Subjects:** Use singular verbs.
  - Example: *The cat* ***runs*** *quickly.*
- **Plural Subjects:** Use plural verbs.
  - Example: *The cats* ***run*** *quickly.*

Common Pitfalls:

- **Intervening Phrases:** Sometimes, words or phrases between the subject and verb can make it harder to see the correct agreement. Always match the verb to the main subject, not to words in the intervening phrase.
  - Incorrect: *The bouquet of flowers* ***are*** *beautiful.*
  - Correct: *The bouquet of flowers* ***is*** *beautiful.*
- **Subjects Joined by "Or" or "Nor":** When subjects are joined by "or" or "nor," the verb should agree with the subject closest to it.
  - Example: *Neither the teacher nor the students* ***were*** *prepared for the test.*
  - Example: *Neither the students nor the teacher* ***was*** *prepared for the test.*

Practice Tip: When faced with subject-verb agreement questions, always identify the subject of the sentence first. Then, determine if it is singular or plural and ensure that the verb matches.

Verb Tense Consistency

Verb tense consistency is crucial in maintaining clarity and coherence in your writing. This rule ensures that the tense remains the same throughout a passage or within a single sentence unless a shift in time is explicitly required.

Basic Rules:

- **Past Tense:** Used for actions that have already occurred.
  - Example: *She **walked** to the store yesterday.*
- **Present Tense:** Used for actions happening now or regularly.
  - Example: *She **walks** to the store every day.*
- **Future Tense:** Used for actions that will occur.
  - Example: *She **will walk** to the store tomorrow.*

Common Pitfalls:

- **Unnecessary Tense Shifts:** Shifting tenses without a clear reason can confuse the reader and disrupt the flow of the writing.
  - Incorrect: *She **walks** to the store and **bought** groceries.*
  - Correct: *She **walked** to the store and **bought** groceries.*
- **Tense in Dependent Clauses:** Ensure that dependent clauses are consistent with the tense of the main clause.
  - Incorrect: *She said that she **will go** to the store.*
  - Correct: *She said that she **would go** to the store.*

Practice Tip: When answering verb tense questions, read the entire sentence or paragraph to understand the timeline of events. Ensure that all verbs are in the correct tense relative to one another.

Pronoun Usage

Pronouns must agree with the nouns they replace in number (singular or plural), gender, and person. Clear antecedent identification is also crucial to avoid confusion.

Basic Rules:

- **Agreement in Number:** Singular nouns take singular pronouns, and plural nouns take plural pronouns.
  - Example: *Each student must bring **his or her** notebook.*
  - Example: *All students must bring **their** notebooks.*

- **Clear Antecedents:** The noun to which a pronoun refers (the antecedent) should be clear and unambiguous.
  - Incorrect: *When Jessica and Amanda arrived, she was tired.*
  - Correct: *When Jessica and Amanda arrived, Jessica was tired.*

Common Pitfalls:

- **Vague Pronouns:** Avoid using pronouns like "it" or "they" without a clear antecedent.
  - Incorrect: *In the novel,* ***it*** *says that...*
  - Correct: *In the novel, the author says that...*
- **Pronoun Case:** Use the correct pronoun case depending on the pronoun's function in the sentence (subject, object, or possessive).
  - Example (Subject): ***He*** *and I went to the store.*
  - Example (Object): *The teacher called* ***him*** *and me to the office.*

Practice Tip: When you encounter a pronoun question, first identify the antecedent and ensure that the pronoun agrees in number, gender, and clarity.

Parallel Structure

Parallel structure, or parallelism, refers to using the same grammatical form for items in a list or connected by coordinating conjunctions (like "and," "or," or "but"). It ensures that similar ideas are expressed in a consistent and balanced way.

Basic Rules:

- **In Lists:** Items in a list should be in the same grammatical form.
  - Incorrect: *She enjoys reading, to jog, and cooking.*
  - Correct: *She enjoys reading, jogging, and cooking.*
- **In Comparisons:** Items being compared should follow the same structure.
  - Incorrect: *He is more interested in reading than to write.*
  - Correct: *He is more interested in reading than in writing.*

Common Pitfalls:

- **Mixed Structures:** Avoid mixing different grammatical forms in a sentence or list.
  - Incorrect: *The course covers grammar, writing, and how to research.*
  - Correct: *The course covers grammar, writing, and researching.*

Practice Tip: When you see a list or comparison in a question, check to ensure that each item follows the same grammatical pattern. Consistency is key to maintaining parallel structure.

Modifier Placement

Modifiers (words, phrases, or clauses that describe something else in the sentence) should be placed as close as possible to the word they modify. Misplaced or dangling modifiers can lead to confusion or unintended meanings.

Basic Rules:

- **Misplaced Modifiers:** Place the modifier next to the word it modifies.
  - Incorrect: *She served pancakes to the children on paper plates.*
  - Correct: *She served pancakes on paper plates to the children.*
- **Dangling Modifiers:** Ensure that the word being modified is clearly stated in the sentence.
  - Incorrect: *Running through the park, the trees were beautiful.*
  - Correct: *Running through the park, I admired the beautiful trees.*

Common Pitfalls:

- **Modifiers at the Beginning of a Sentence:** When starting a sentence with a modifying phrase, make sure the subject of the sentence is what the modifier describes.
  - Incorrect: *While driving to work, the traffic was heavy.*
  - Correct: *While driving to work, I noticed that the traffic was heavy.*

Practice Tip: Read sentences carefully to ensure that the modifier is placed correctly. If a sentence feels awkward or unclear, check for misplaced or dangling modifiers.

Practice Questions and Explanations

To solidify your understanding of these grammar and usage rules, practice with the following questions:

Practice Question 1: Identify the error in the following sentence. *The team of engineers have completed their project ahead of schedule.*

- A) The team of engineers
- B) have
- C) their
- D) ahead of schedule

Correct Answer: B) *have* (The correct verb should be "has" to agree with the singular subject "team.")

Practice Question 2: Choose the correct sentence.

- A) Neither of the students were prepared for the exam.
- B) Neither of the students was prepared for the exam.

Correct Answer: B) *Neither of the students was prepared for the exam.* (The subject "Neither" is singular and requires a singular verb "was.")

Practice Question 3: Identify the correct sentence.

- A) After the storm, walking through the park, the trees were damaged.
- B) After the storm, I noticed the damaged trees while walking through the park.

Correct Answer: B) *After the storm, I noticed the damaged trees while walking through the park.* (The modifier is correctly placed to clarify who is doing the walking.)

By mastering these grammar and usage rules, you will be better equipped to tackle the Writing and Language Test on the SAT. Regular practice will help reinforce these concepts and improve your ability to identify and correct errors in writing. In the next section, we will explore strategies for effective writing, including sentence structure, organization, and style.

Effective Writing Strategies

The Writing and Language Test not only assesses your grasp of grammar and usage but also evaluates your ability to improve the clarity, coherence, and effectiveness of written passages. This section focuses on strategies that will help you enhance sentence structure, organization, and overall expression in your writing. Mastering these strategies will allow you to approach the Writing and Language Test with confidence and precision.

Sentence Structure and Clarity

Effective writing requires clear and well-constructed sentences. Understanding how to structure sentences for clarity and impact is essential for conveying your ideas effectively.

Key Strategies:

1. **Avoiding Run-on Sentences:**
    - Run-on sentences occur when two or more independent clauses are joined without proper punctuation or conjunctions.
    - **Incorrect:** *The experiment was successful it demonstrated the hypothesis was correct.*
    - **Correct:** *The experiment was successful, and it demonstrated that the hypothesis was correct.*

Tip: Use a comma and a coordinating conjunction (for, and, nor, but, or, yet, so) to join two independent clauses, or use a semicolon to link closely related ideas.

2. **Eliminating Sentence Fragments:**
    - A sentence fragment is an incomplete sentence that lacks a subject, verb, or complete thought.
    - **Incorrect:** *Because the data was inconclusive.*
    - **Correct:** *The study was inconclusive because the data was insufficient.*

Tip: Ensure that every sentence has a subject and a verb and expresses a complete thought.

3. **Using Parallel Structure:**
    - Parallel structure involves using the same grammatical form for related ideas within a sentence. This technique enhances readability and emphasizes the relationship between ideas.
    - **Incorrect:** *She likes hiking, to swim, and running.*
    - **Correct:** *She likes hiking, swimming, and running.*

Tip: When listing items or pairing ideas, ensure that each element follows the same grammatical structure.

4. **Avoiding Wordiness and Redundancy:**

- Wordiness occurs when unnecessary words or phrases are used, which can make sentences cumbersome and unclear.
- **Incorrect:** *In order to fully complete the assignment, she needs to finish it by the due date.*
- **Correct:** *She needs to finish the assignment by the due date.*

Tip: Remove redundant words and phrases to make your writing more concise and direct.

5. **Correcting Misplaced and Dangling Modifiers:**
   - Modifiers should be placed next to the word they describe. A misplaced modifier can create confusion, while a dangling modifier lacks a clear subject.
   - **Incorrect:** *Walking to the store, the flowers caught my eye.*
   - **Correct:** *Walking to the store, I noticed the flowers.*

Tip: Place modifiers close to the word they modify and ensure that the subject of the sentence is clearly stated.

Organization and Coherence

Organizing your writing logically and ensuring that ideas flow smoothly from one to the next is crucial for effective communication. This aspect of writing is heavily tested in the Writing and Language Test.

Key Strategies:

1. **Using Transitional Words and Phrases:**
   - Transitions help guide the reader through your writing by connecting ideas and signaling relationships between them.
   - **Example:** *Moreover, the study revealed significant results, which further support the initial hypothesis.*

Tip: Use transitions like "however," "therefore," "in addition," and "for example" to clarify the connections between your ideas.

2. **Maintaining a Logical Flow of Ideas:**
   - Ensure that each sentence and paragraph follow logically from the one before it. This creates a coherent argument or narrative.
   - **Example:** *First, the team gathered data. Next, they analyzed the results. Finally, they presented their findings.*

Tip: Before writing, outline your ideas to ensure a logical progression from introduction to conclusion.

3. **Developing Strong Topic Sentences:**
   - A topic sentence introduces the main idea of a paragraph and sets the tone for the sentences that follow.
   - **Example:** *The rise of renewable energy sources is transforming the global energy market.*

Tip: Each paragraph should begin with a topic sentence that clearly states the main idea, followed by supporting details.

4. **Ensuring Paragraph Unity:**

- A paragraph should focus on a single idea, with all sentences contributing to the development of that idea.
- **Example:** *The advantages of solar power are numerous. It is a renewable energy source, it reduces greenhouse gas emissions, and it has the potential to lower energy costs.*

Tip: Avoid introducing new ideas in the middle of a paragraph. Stick to the topic introduced by the topic sentence.

5. **Refining Introductions and Conclusions:**
   - The introduction should provide a clear overview of the topic, while the conclusion should summarize the key points and provide a sense of closure.
   - **Example (Introduction):** *Climate change is one of the most pressing issues of our time, affecting ecosystems and human societies worldwide.*
   - **Example (Conclusion):** *In conclusion, addressing climate change requires global cooperation, innovative solutions, and a commitment to sustainability.*

Tip: Revisit your introduction and conclusion after writing to ensure they effectively frame your main argument or narrative.

Style and Tone

The Writing and Language Test also assesses your ability to match the style and tone of a passage to its intended audience and purpose. Writing that is appropriate in style and tone is clear, engaging, and suitable for the context.

Key Strategies:

1. **Adjusting Style for Audience and Purpose:**
   - Formal writing is appropriate for academic and professional contexts, while informal writing may be suitable for casual or personal communication.
   - **Formal Example:** *The study's findings indicate a significant correlation between diet and health outcomes.*
   - **Informal Example:** *So, it turns out that what you eat really does matter.*

Tip: Consider the context of the passage and adjust your word choice, sentence structure, and tone accordingly.

2. **Using Precise Language:**
   - Choose words that convey your meaning as clearly and accurately as possible. Avoid vague or ambiguous language.
   - **Vague:** *The experiment was kind of successful.*
   - **Precise:** *The experiment yielded positive results.*

Tip: Opt for specific terms that leave no room for misinterpretation.

3. **Maintaining Consistent Tone:**

- The tone of a passage should remain consistent throughout. Shifts in tone can confuse the reader or disrupt the flow of the writing.
- **Inconsistent:** *The results were promising, but unfortunately, they weren't good enough to continue.*
- **Consistent:** *The results were promising, but they did not meet the criteria for further investigation.*

Tip: Consider the overall mood or attitude conveyed in the passage and ensure that it is maintained from start to finish.

4. **Avoiding Overly Complex Language:**
   - While sophisticated vocabulary can enhance your writing, overly complex language can obscure your meaning and alienate the reader.
   - **Overly Complex:** *The amelioration of the atmospheric conditions facilitated the expedition.*
   - **Simpler Alternative:** *The improved weather conditions made the expedition easier.*

Tip: Strive for clarity and simplicity, ensuring that your writing is accessible and understandable.

Practice Questions and Explanations

To reinforce these strategies, practice with the following questions:

Practice Question 1: Identify the best revision for the following sentence. *The committee members discussed the issue for hours, but ultimately, they didn't reach any conclusion that everyone agreed on.*

- A) The committee members discussed the issue for hours, but ultimately, they didn't reach a conclusion that everyone agreed on.
- B) The committee members discussed the issue for hours, but ultimately, they didn't reach a conclusion on which everyone agreed.
- C) The committee members discussed the issue for hours, but ultimately, they didn't reach any conclusions that everyone could agree with.
- D) The committee members discussed the issue for hours, but ultimately, they didn't reach a conclusion that everyone could agree with.

Correct Answer: B) *The committee members discussed the issue for hours, but ultimately, they didn't reach a conclusion on which everyone agreed.* (This option maintains formal tone and clarity.)

Practice Question 2: Choose the sentence that best maintains parallel structure.

- A) The program aims to improve efficiency, increase productivity, and to reduce costs.
- B) The program aims to improve efficiency, increasing productivity, and reducing costs.
- C) The program aims to improve efficiency, increase productivity, and reduce costs.
- D) The program aims to improve efficiency, and to increase productivity, and reduce costs.

Correct Answer: C) *The program aims to improve efficiency, increase productivity, and reduce costs.* (This option maintains parallel structure.)

Practice Question 3: Which of the following sentences is most concise and clear?

- A) The data provided by the experiment was somewhat inconclusive due to a variety of factors that weren't accounted for in the initial planning stages.
- B) The data from the experiment was inconclusive due to unaccounted factors in the planning stages.
- C) The experiment's data was inconclusive because of unaccounted factors in the initial planning.
- D) The data was inconclusive due to factors not accounted for in the planning.

Correct Answer: D) *The data was inconclusive due to factors not accounted for in the planning.* (This option is concise and clear.)

By mastering these effective writing strategies, you will be well-prepared to enhance and refine passages on the SAT Writing and Language Test. These strategies, combined with regular practice, will help you improve your score and build your confidence in writing and editing. In the next section, we will focus on vocabulary in context, an essential skill for both.

Vocabulary in Context

Vocabulary in Context is a critical skill tested in both the Reading and Writing sections of the SAT. This skill involves understanding the meaning of words and phrases as they are used in specific contexts, rather than relying solely on dictionary definitions. Mastering vocabulary in context allows you to comprehend passages more accurately and choose the most appropriate words to enhance clarity and precision in writing.

Understanding Vocabulary in Context

The SAT frequently asks you to determine the meaning of a word or phrase based on its context within a passage. This means that even if you're unfamiliar with a particular word, you can often infer its meaning by analyzing the surrounding text.

Key Strategies:

1. **Use Context Clues:**
   - **Definition Clues:** Sometimes, the passage directly defines the word.
     - Example: *The biologist discovered an* ***indigenous*** *species, meaning it was native to the region.*
   - **Synonym Clues:** A word with a similar meaning may be used nearby.
     - Example: *The city was in a state of* ***turmoil****, or disorder, after the storm.*
   - **Antonym Clues:** A word with the opposite meaning can also provide hints.
     - Example: *Unlike the* ***peripheral*** *buildings, which were on the outskirts, the central plaza was well maintained.*
   - **Explanation Clues:** The author may explain the meaning of the word through further elaboration.
     - Example: *The decision was* ***contentious****, sparking heated debate among the council members.*

Tip: When you encounter an unfamiliar word, look at the sentences before and after it. These can provide valuable context clues.

2. **Consider Word Roots and Affixes:**
   - Understanding the root of a word, as well as common prefixes and suffixes, can help you deduce its meaning.
     - Example: The word **"benevolent"** comes from the Latin root *bene*, meaning "good" or "well," and *volent*, from *volens*, meaning "wishing." Together, they suggest a meaning related to kindness or goodwill.

Tip: Familiarize yourself with common word roots and affixes, as these can often lead you to the correct meaning of a word.

3. **Analyze the Function of the Word in the Sentence:**
    - Consider how the word is used in the sentence. Is it functioning as a noun, verb, adjective, or adverb? This can provide additional clues to its meaning.
        - Example: In the sentence *"He gave a **cursory** glance at the report,"* the word **cursory** functions as an adjective describing the type of glance, implying it was quick and not thorough.

Tip: Identifying the part of speech can narrow down the possible meanings of the word.

4. **Avoid Overreliance on Familiar Meanings:**
    - Some words have multiple meanings, and the SAT may test a less common usage of a familiar word. Ensure that the meaning you choose fits the context of the passage.
        - Example: The word **"table"** can mean both a piece of furniture and a verb meaning to postpone discussion of something.

Tip: Always re-read the sentence to ensure that the meaning you select makes sense in the given context.

Building a Strong Vocabulary

While context clues can help you infer the meaning of unfamiliar words, having a strong vocabulary will make this task easier and quicker. Here are strategies for building and strengthening your vocabulary:

1. **Read Regularly:**
    - Exposure to a wide range of texts will naturally expand your vocabulary. Read books, articles, and essays from various genres and disciplines.
    - **Tip:** Keep a list of unfamiliar words you encounter during reading and look up their meanings. Review this list regularly to reinforce your understanding.
2. **Use Vocabulary Lists:**
    - Many SAT prep books and online resources provide lists of commonly tested vocabulary words. These lists are a great starting point for building a strong vocabulary.
    - **Tip:** Create flashcards for these words, including their definitions and an example sentence. Quiz yourself regularly.
3. **Practice Using New Words:**
    - Incorporate new vocabulary into your everyday writing and speaking. The more you use a word, the more likely you are to remember it.
    - **Tip:** Try writing sentences or short paragraphs using several new words to practice their usage in context.
4. **Engage with Word Games and Apps:**
    - Word games and vocabulary-building apps can make learning new words enjoyable and effective.
    - **Tip:** Apps like Quizlet, Memrise, and others are designed to help you learn and retain new vocabulary through interactive exercises.

Practice Questions and Explanations

To solidify your understanding of vocabulary in context, practice with the following questions:

Practice Question 1: Based on the context, what does the word "arduous" most nearly mean in the sentence below? *The climb to the summit was arduous, but the breathtaking view made it worthwhile.*

- A) Easy
- B) Exciting
- C) Difficult
- D) Quick

Correct Answer: C) *Difficult.* (The context of the climb being challenging implies that "arduous" means difficult.)

Practice Question 2: In the passage, the word "mitigate" most nearly means: *The city council introduced measures to mitigate the effects of the economic downturn.*

- A) Intensify
- B) Alleviate
- C) Ignore
- D) Emphasize

Correct Answer: B) *Alleviate.* (The context suggests that the measures were meant to lessen the negative effects of the economic downturn.)

Practice Question 3: The word "prolific" as used in the sentence below most nearly means: *The author was prolific, publishing dozens of novels throughout her career.*

- A) Lazy
- B) Limited
- C) Productive
- D) Unsuccessful

Correct Answer: C) *Productive.* (The context of publishing many novels indicates that "prolific" means highly productive.)

By mastering vocabulary in context, you will be better equipped to understand and analyze the passages on the SAT Reading and Writing sections. A strong vocabulary not only improves your reading comprehension but also enhances your ability to write clearly and effectively. In the next section, we will provide practice passages and questions to further hone your skills and prepare you for test day.

Practice Passages and Questions

The best way to prepare for the Reading and Writing sections of the SAT is through consistent practice with passages that mirror the style and content of the actual test. In this section, you will find practice passages followed by multiple-choice questions designed to test your comprehension, analysis, and editing skills. Detailed explanations for each question will help you understand why certain answers are correct and others are not, further reinforcing the strategies you've learned.

Practice Passage 1: Literature

Passage: *This passage is adapted from a novel that explores the complexities of family relationships and personal growth.*
"Maria stared out the window, watching as the rain pattered gently against the glass. It had been three years since she had last visited her childhood home, and the sight of the familiar landscape, now shrouded in mist, brought a wave of nostalgia. The once-vibrant garden, now overgrown and wild, mirrored her own feelings of abandonment. She had left this place, and the people in it, to chase a dream that now seemed so distant. But here she was, back again, with a heart full of questions and a mind weary of the answers she might find."

Questions:

1. What is the main idea of the passage?

- A) Maria is excited to return to her childhood home.
- B) Maria feels a sense of abandonment and uncertainty upon returning home.
- C) The passage describes a vibrant and well-maintained garden.
- D) Maria reflects on her successful journey and feels content.

2. The phrase "with a heart full of questions" most likely suggests that Maria:

- A) is curious about the changes in her childhood home.
- B) is unsure about her decision to return home.
- C) is eager to reconnect with her family.
- D) is confident about the answers she will find.

3. In the context of the passage, the word "weary" most nearly means:

- A) enthusiastic
- B) tired
- C) indifferent
- D) relieved

Answers and Explanations:

1. **Correct Answer:** B) *Maria feels a sense of abandonment and uncertainty upon returning home.* (The passage conveys Maria's feelings of nostalgia and doubt, suggesting she is grappling with her emotions as she returns home.)
2. **Correct Answer:** B) *is unsure about her decision to return home.* (The phrase "heart full of questions" implies that Maria is experiencing doubt and uncertainty about what she will discover upon her return.)
3. **Correct Answer:** B) *tired.* (In this context, "weary" describes Maria's emotional state, indicating that she is mentally and emotionally exhausted.)

Practice Passage 2: History/Social Studies

Passage: *This passage is adapted from a speech delivered in the early 20th century by a prominent leader advocating for women's suffrage.*

"We stand at a pivotal moment in history, where the voices of women, long silenced, are finally beginning to be heard. The fight for suffrage is not merely a struggle for the vote; it is a fight for recognition, for equality, for the acknowledgment that women, like men, possess the intellect, the moral fortitude, and the right to participate fully in the governance of this nation. It is a fight that transcends borders and speaks to the universal truth that all human beings are entitled to the same rights and freedoms. We must not falter in our efforts, for the future of our daughters and granddaughters depends on the progress we make today."

Questions:

1. The main purpose of the passage is to:

- A) describe the historical context of women's suffrage.
- B) argue that women should have the right to vote.
- C) explain the differences between men's and women's rights.
- D) discuss the challenges faced by women in the 20th century.

2. In the context of the passage, what does the word "fortitude" most nearly mean?

- A) physical strength
- B) resilience
- C) intelligence
- D) compassion

3. The phrase "transcends borders" most likely means that the fight for women's suffrage:

- A) is limited to one country.
- B) is relevant only in the 20th century.
- C) applies to all people, regardless of nationality.
- D) requires physical travel across nations.

Answers and Explanations:

1. **Correct Answer:** B) *argue that women should have the right to vote.* (The passage is a persuasive speech advocating for women's suffrage, emphasizing the importance of equal rights.)
2. **Correct Answer:** B) *resilience.* (The word "fortitude" in this context refers to the strength and resilience required to endure the fight for women's rights.)
3. **Correct Answer:** C) *applies to all people, regardless of nationality.* (The phrase "transcends borders" suggests that the issue of women's suffrage is a universal one, not confined to a single nation.)

Practice Passage 3: Science

Passage: *This passage is adapted from a scientific article discussing recent advancements in renewable energy technology.*

"In recent years, the development of renewable energy sources has accelerated, driven by the urgent need to reduce carbon emissions and combat climate change. Among these sources, solar power has emerged as one of the most promising. Advances in photovoltaic cell technology have significantly increased the efficiency of solar panels, making them more cost-effective and accessible than ever before. Moreover, the integration of energy storage systems has addressed one of the primary challenges associated with solar power: its intermittent nature. As researchers continue to innovate, the potential for solar energy to become a dominant global power source grows ever more likely."

Questions:

1. The passage is primarily focused on:

- A) the history of solar power development.
- B) the environmental impact of solar energy.
- C) recent advancements in solar power technology.
- D) the cost of implementing solar power globally.

2. In the context of the passage, the word "intermittent" most nearly means:

- A) constant
- B) unreliable
- C) occasional
- D) predictable

3. What is one of the challenges mentioned in the passage associated with solar power?

- A) High production costs
- B) Limited availability of materials
- C) Intermittent energy generation
- D) Environmental degradation

Answers and Explanations:

1. **Correct Answer:** C) *recent advancements in solar power technology.* (The passage focuses on technological improvements in solar energy, highlighting increased efficiency and the potential for growth.)
2. **Correct Answer:** C) *occasional.* (The word "intermittent" refers to something that occurs at irregular intervals, which fits the context of the passage describing the inconsistent availability of solar energy.)

3. **Correct Answer:** C) *Intermittent energy generation.* (The passage mentions the challenge of solar power's intermittent nature, which has been addressed through energy storage systems.)

Practice Passage 4: Humanities

Passage: *This passage is adapted from an essay on the role of art in society.*

"Art has always served as a reflection of the human experience, capturing the joys, sorrows, triumphs, and tragedies that define our existence. From the earliest cave paintings to modern digital installations, art has been a medium through which individuals express their deepest emotions and convey messages that resonate across time and culture. In a world increasingly driven by technology and data, the role of art as a conduit for empathy and understanding has never been more crucial. Art challenges us to see the world through different perspectives, to question our assumptions, and to connect with one another on a profoundly human level."

Questions:

1. The author of the passage suggests that the primary role of art is to:

- A) entertain and amuse audiences.
- B) provide a historical record of events.
- C) serve as a reflection of human experiences and emotions.
- D) advance technological innovation.

2. The phrase "conduit for empathy" most likely means that art:

- A) helps people feel compassion for others.
- B) serves as a distraction from daily life.
- C) focuses solely on negative emotions.
- D) is a means of technical communication.

3. According to the passage, art challenges us to:

- A) accept traditional viewpoints.
- B) embrace modern technology.
- C) question our assumptions and broaden our perspectives.
- D) avoid emotional connections with others.

Answers and Explanations:

1. **Correct Answer:** C) *serve as a reflection of human experiences and emotions.* (The passage emphasizes art's role in expressing emotions and experiences that resonate with people across cultures and time.)
2. **Correct Answer:** A) *helps people feel compassion for others.* (The phrase "conduit for empathy" suggests that art fosters understanding and compassion among people.)
3. **Correct Answer:** C) *question our assumptions and broaden our perspectives.* (The passage highlights art's ability to challenge our views and encourage us to see the world in new ways.)

By practicing with these passages and questions, you can strengthen your reading comprehension, analytical skills, and your ability to edit and improve written material. Regular practice will not only familiarize you with the types of questions you'll encounter on the SAT but also build the confidence you need to perform well on test day. In the next section, we will focus on test-taking strategies specifically tailored for the Evidence-Based Reading and Writing section, helping you maximize your score through effective techniques.

Test-Taking Strategies for Evidence-Based Reading and Writing

Success on the SAT's Evidence-Based Reading and Writing section requires not just knowledge and skills, but also effective test-taking strategies. Understanding how to manage your time, approach different types of questions, and maintain focus during the test can significantly enhance your performance. This section provides practical strategies that will help you maximize your score.

General Test-Taking Strategies

1. **Pace Yourself:**
   - Time management is crucial on the SAT. For the Reading Test, you have 65 minutes to answer 52 questions, which averages about 75 seconds per question. For the Writing and Language Test, you have 35 minutes to answer 44 questions, giving you about 48 seconds per question. Be mindful of the clock, but don't rush through questions.
   - **Tip:** Use your first pass to answer questions you find easy, then return to more challenging questions with the remaining time.
2. **Answer Every Question:**
   - There is no penalty for guessing on the SAT, so it's in your best interest to answer every question, even if you're unsure of the correct answer.
   - **Tip:** If you're running out of time, make educated guesses on remaining questions rather than leaving them blank.
3. **Eliminate Wrong Answers:**
   - For multiple-choice questions, use the process of elimination to narrow down your choices. Eliminating even one or two incorrect options increases your chances of selecting the correct answer.
   - **Tip:** If an answer choice doesn't align with the main idea of the passage or seems out of context, it's likely incorrect.
4. **Stay Focused and Calm:**
   - Maintaining focus during the test is critical. If you find yourself getting anxious or losing concentration, take a few deep breaths and refocus on the task at hand.
   - **Tip:** Break the test into smaller segments in your mind (e.g., focus on one passage at a time) to avoid feeling overwhelmed.
5. **Use Context to Your Advantage:**
   - For vocabulary in context questions, the surrounding sentences often provide clues to the meaning of an unfamiliar word. Pay attention to how the word is used in the passage.
   - **Tip:** Re-read the sentence with each answer choice to see which one makes the most sense in the context of the passage.

Strategies for the Reading Test

1. **Skim the Passage First:**
    - Before diving into the questions, quickly skim the passage to get a general sense of its content, structure, and tone. This can help you answer questions more efficiently.
    - **Tip:** Look for key elements such as the main idea, author's purpose, and overall structure during your initial read.
2. **Identify the Main Idea and Purpose:**
    - Many questions on the Reading Test revolve around the main idea or purpose of the passage. Pay attention to the thesis statement or summary sentences that highlight the central argument.
    - **Tip:** If a question asks about the author's purpose, consider why the passage was written (e.g., to inform, persuade, entertain).
3. **Tackle Passage-Specific Questions First:**
    - Questions that refer to specific lines or details in the passage can often be answered more quickly than broader, thematic questions. Answer these first to build confidence and save time for more complex questions.
    - **Tip:** Use the line references in the questions to locate relevant information quickly.
4. **Synthesize Information in Paired Passages:**
    - When dealing with paired passages, focus on understanding each passage individually before answering questions that compare the two. Pay attention to the relationship between the passages (e.g., agree/disagree, different perspectives on the same issue).
    - **Tip:** Answer questions about each passage separately before attempting synthesis questions that require comparing both.
5. **Use Active Reading Techniques:**
    - Mark up the passage as you read by underlining key points, circling transition words, and jotting down brief notes in the margins. Active reading helps you stay engaged and better retain information.
    - **Tip:** Focus on identifying the author's argument, evidence, and any shifts in tone or perspective.

Strategies for the Writing and Language Test

1. **Focus on Clarity and Precision:**
    - The Writing and Language Test is all about improving the clarity and effectiveness of the passages. When answering questions, prioritize revisions that make the text clearer, more concise, and easier to understand.
    - **Tip:** Choose answers that eliminate redundancy, wordiness, and ambiguity.
2. **Check for Consistency:**
    - Ensure that verb tenses, pronouns, and tone are consistent throughout the passage. Inconsistent language can confuse the reader and disrupt the flow of the writing.
    - **Tip:** When you encounter a question about verb tense, scan the surrounding sentences to confirm the correct tense.
3. **Match the Tone and Style:**
    - Some questions will ask you to choose the word or phrase that best matches the tone and style of the passage. Consider the audience and purpose of the text when selecting your answer.

- **Tip:** If the passage is formal, avoid informal language in your revisions, and vice versa.

4. **Use the Process of Elimination:**
    - For grammar and usage questions, eliminate answer choices that introduce errors or awkward phrasing. The correct answer should improve the passage without changing its intended meaning.
    - **Tip:** Read each answer choice in the context of the sentence to determine which one fits best.
5. **Prioritize Sentence Structure and Flow:**
    - Questions about sentence structure often involve improving the logical flow of ideas. Choose revisions that enhance the passage's organization and coherence.
    - **Tip:** Look for transition words that indicate relationships between ideas (e.g., however, therefore, additionally).
6. **Look for Common Grammatical Errors:**
    - The Writing and Language Test frequently includes questions on subject-verb agreement, pronoun clarity, and parallel structure. Be on the lookout for these common pitfalls.
    - **Tip:** When you spot a potential grammatical error, double-check that the subject and verb agree, pronouns have clear antecedents, and parallel structures are maintained.

Practice Scenarios and Timing Strategies

To ensure that you're fully prepared, it's essential to practice with timed scenarios that simulate the actual test environment. Here are some tips for effective practice:

1. **Simulate Test Conditions:**
    - Take full-length practice tests under timed conditions to build stamina and familiarity with the test format. This will help you gauge your pacing and identify areas where you may need to improve.
    - **Tip:** Set aside uninterrupted time, use a timer, and avoid distractions to mimic real test conditions.
2. **Review Your Mistakes:**
    - After completing a practice test, review every question you got wrong or found difficult. Understanding why an answer is incorrect is just as important as knowing the correct answer.
    - **Tip:** Keep a journal of common mistakes and review it regularly to avoid repeating them on test day.
3. **Work on Your Weaknesses:**
    - Focus on the areas where you struggle the most. Whether it's vocabulary in context, sentence structure, or reading comprehension, targeted practice can help you improve specific skills.
    - **Tip:** Use online resources, tutoring, or study groups to get extra help in your weaker areas.
4. **Develop a Personal Timing Strategy:**
    - Everyone's test-taking style is different. Some students may benefit from spending more time on the passage reading and less on the questions, while others may prefer the opposite. Develop a timing strategy that works best for you.
    - **Tip:** Experiment with different approaches during practice tests to find the strategy that maximizes your accuracy and speed.

Managing Test Anxiety

Test anxiety can negatively impact your performance if not properly managed. Here are some techniques to help you stay calm and focused:

1. **Practice Relaxation Techniques:**
    - Deep breathing exercises, visualization, and mindfulness can help reduce anxiety before and during the test.
    - **Tip:** Practice these techniques regularly so they become second nature on test day.
2. **Stay Positive:**
    - Maintain a positive mindset and remind yourself that you're well-prepared. Confidence can help reduce stress and improve your performance.
    - **Tip:** Use positive affirmations, such as "I am prepared" and "I can do this," to boost your confidence.
3. **Take Care of Your Physical Health:**
    - A healthy body supports a healthy mind. Ensure you're well-rested, hydrated, and have eaten a balanced meal before the test.
    - **Tip:** Avoid caffeine and sugar, which can increase anxiety and lead to energy crashes during the test.
4. **Have a Plan for Test Day:**
    - Knowing what to expect on test day can help reduce anxiety. Plan your route to the test center, know what to bring, and arrive early to avoid last-minute stress.
    - **Tip:** Prepare your test-day materials (e.g., admission ticket, ID, calculator) the night before to ensure you're ready to go.

By applying these test-taking strategies, you can approach the Evidence-Based Reading and Writing section with confidence and focus. Remember that consistent practice, effective time management, and a calm mindset are key to achieving your best possible score. In the next section, we will explore full-length practice tests that allow you to put these strategies into action and assess your progress.

Review and Practice Tests

After learning the strategies and skills necessary to excel in the Evidence-Based Reading and Writing section of the SAT, it's crucial to put everything into practice. This section provides full-length practice tests that simulate the actual SAT experience, allowing you to apply what you've learned, gauge your performance, and identify areas where you may need further improvement.

Importance of Practice Tests

Taking full-length practice tests is one of the most effective ways to prepare for the SAT. These tests help you:

1. **Build Stamina:** The SAT is a lengthy exam, and maintaining focus and energy throughout is essential. Practice tests help you develop the endurance needed to perform well on test day.
2. **Refine Timing:** Timing is critical on the SAT, and practice tests allow you to hone your time management skills. By practicing under timed conditions, you can learn to pace yourself appropriately and avoid spending too much time on any one question.
3. **Identify Strengths and Weaknesses:** Reviewing your performance on practice tests helps you pinpoint areas where you excel and areas where you need additional practice. This targeted approach ensures that you spend your study time effectively.
4. **Familiarize Yourself with the Test Format:** Regular practice with full-length tests makes you more comfortable with the SAT format, reducing anxiety and boosting confidence on test day.

How to Use This Section

This section includes multiple full-length practice tests for the Evidence-Based Reading and Writing section. Each test is designed to replicate the content, structure, and timing of the actual SAT. After completing a practice test, you'll find detailed answer explanations and scoring guides to help you assess your performance.

Steps to Follow:

1. **Simulate Real Test Conditions:**
    - Find a quiet space where you won't be interrupted.
    - Set a timer according to the official SAT time limits (65 minutes for the Reading Test, 35 minutes for the Writing and Language Test).
    - Complete the entire test in one sitting to replicate the experience of the actual exam.

2. **Take the Test:**
   - Work through the questions methodically, applying the strategies you've learned.
   - Answer every question, even if you have to guess on some.
   - Keep an eye on the time, but don't rush. Balance speed with accuracy.
3. **Score Your Test:**
   - Use the provided answer key to score your test. Mark each correct answer, and calculate your raw score for both the Reading and Writing sections.
4. **Review Answer Explanations:**
   - For each question you answered incorrectly, read the detailed explanation provided. Understanding why an answer is correct or incorrect is crucial for improving your performance.
   - Pay attention to any patterns in your mistakes. Are there specific types of questions you find challenging? Are there particular skills or strategies you need to work on?
5. **Analyze Your Performance:**
   - After reviewing the explanations, reflect on your overall performance. Consider the following:
     - Did you manage your time effectively?
     - Were there any distractions that affected your concentration?
     - Did you struggle with certain passages or question types?
   - Use this analysis to guide your future study sessions.
6. **Set Goals for Improvement:**
   - Based on your performance, set specific, measurable goals for your next practice test. For example:
     - Improve your score on vocabulary in context questions.
     - Reduce the time spent on each passage.
     - Increase your accuracy in answering inference questions.
   - Plan your study schedule to focus on achieving these goals.

Full-Length Practice Test

Reading Test - Passage 1: Literature

This passage is adapted from Jane Austen's "Pride and Prejudice," originally published in 1813. In this scene, Elizabeth Bennet visits Pemberley, the estate of Mr. Darcy, for the first time.

Elizabeth, as they drove into the park, was delighted with its beauty; and when they passed through the gates, she thought she had never seen a place for which nature had done more, or where natural beauty had been so little counteracted by an awkward taste. They were all of them warm in their admiration; and at that moment she felt that to be mistress of Pemberley might be something!

They gradually ascended for half-a-mile, and then found themselves at the top of a considerable eminence, where the wood ceased, and the eye was instantly caught by Pemberley House, situated on the opposite side of a valley, into which the road, with some abruptness, wound. It was a large, handsome stone building, standing well on rising ground, and backed by a ridge of high woody hills;—and in front, a stream of some natural importance was swelled into greater, but without any artificial appearance. Its banks were neither formal nor falsely adorned.

Elizabeth was too much excited to speak; and at length the carriage stopped. The housekeeper came to meet them at the door, and as they walked through the hall, into the dining-parlour, Elizabeth found herself with less inclination for company than ever. The rooms were lofty and handsome, and their furniture suitable to the fortune of its proprietor; but Elizabeth saw, with admiration of his taste, that it was neither gaudy nor uselessly fine; with less of splendour, and more real elegance, than the furniture of Rosings.

Questions:

1. What is the main impression Elizabeth has of Pemberley upon her arrival?
   - A) She finds it too extravagant for her taste.
   - B) She feels that it is a perfect blend of natural beauty and human design.
   - C) She is disappointed by its lack of grandeur.
   - D) She is overwhelmed by the artificial beauty of the estate.
2. The phrase "natural beauty had been so little counteracted by an awkward taste" most nearly means:
   - A) The natural landscape had been poorly altered by human hands.
   - B) The estate's design enhanced its natural surroundings.

    - C) The estate's beauty was ruined by poor maintenance.
    - D) The natural beauty was completely overshadowed by the house's architecture.
3. What can be inferred about Elizabeth's feelings towards Mr. Darcy from her reaction to Pemberley?
    - A) She is indifferent towards him.
    - B) She sees the estate as a reflection of his good character.
    - C) She believes he has poor taste in design.
    - D) She is critical of the way he has managed the estate.
4. How does the author's description of Pemberley contribute to the overall mood of the passage?
    - A) It creates a sense of awe and admiration.
    - B) It evokes a feeling of unease.
    - C) It introduces a tone of disappointment.
    - D) It highlights the grandeur and formality of the setting.
5. Based on the passage, what is the significance of Elizabeth's comparison of Pemberley to Rosings?
    - A) It shows her preference for simplicity over opulence.
    - B) It indicates her desire to live a life of luxury.
    - C) It suggests that she finds Pemberley inferior to Rosings.
    - D) It reveals her disdain for all forms of wealth.

Answers and Explanations:

1. **Question:** What is the main impression Elizabeth has of Pemberley upon her arrival?
    - **Correct Answer:** B) *She feels that it is a perfect blend of natural beauty and human design.*
    - **Explanation:** The passage emphasizes Elizabeth's admiration for how Pemberley's natural beauty has been preserved and enhanced by tasteful human design. This is evident in her thoughts about the estate's natural beauty not being counteracted by awkward taste.
2. **Question:** The phrase "natural beauty had been so little counteracted by an awkward taste" most nearly means:
    - **Correct Answer:** B) *The estate's design enhanced its natural surroundings.*
    - **Explanation:** This phrase suggests that the natural beauty of Pemberley has not been diminished by poor or excessive human design. Instead, the design complements and enhances the natural beauty, which is why Elizabeth admires it.
3. **Question:** What can be inferred about Elizabeth's feelings towards Mr. Darcy from her reaction to Pemberley?
    - **Correct Answer:** B) *She sees the estate as a reflection of his good character.*
    - **Explanation:** Elizabeth's positive reaction to Pemberley suggests that she views the estate as a reflection of Mr. Darcy's taste and character. Her admiration for the estate implies that she might be reassessing her opinion of Mr. Darcy, seeing him in a more favorable light.
4. **Question:** How does the author's description of Pemberley contribute to the overall mood of the passage?
    - **Correct Answer:** A) *It creates a sense of awe and admiration.*
    - **Explanation:** The detailed and positive description of Pemberley evokes a mood of awe and admiration. Elizabeth is deeply impressed by the estate, which contributes to the overall positive tone of the passage.
5. **Question:** Based on the passage, what is the significance of Elizabeth's comparison of Pemberley to Rosings?
    - **Correct Answer:** A) *It shows her preference for simplicity over opulence.*
    - **Explanation:** Elizabeth notes that Pemberley has "less of splendour, and more real elegance" compared to Rosings. This comparison indicates her preference for elegance and taste over excessive opulence, reflecting her values and character.

Reading Test - Passage 2: History/Social Studies

This passage is adapted from a speech delivered by Frederick Douglass in 1852, titled "What to the Slave is the Fourth of July?" Douglass, a former slave, was a prominent abolitionist and social reformer. In this speech, he discusses the meaning of Independence Day to African Americans during the era of slavery.

Fellow-citizens, pardon me, allow me to ask, why am I called upon to speak here today? What have I, or those I represent, to do with your national independence? Are the great principles of political freedom and of natural justice, embodied in that Declaration of Independence, extended to us? And am I, therefore, called upon to bring our humble offering to the national altar, and to confess the benefits and express devout gratitude for the blessings resulting from your independence to us?

But, such is not the state of the case. I say it with a sad sense of the disparity between us. I am not included within the pale of this glorious anniversary! Your high independence only reveals the immeasurable distance between us. The blessings in which you, this day, rejoice, are not enjoyed in common. The rich inheritance of justice, liberty, prosperity, and independence, bequeathed by your fathers, is shared by you, not by me. The sunlight that brought life and healing to you has brought stripes and death to me. This Fourth [of] July is yours, not mine. You may rejoice, I must mourn. To drag a man in fetters into the grand illuminated temple of liberty, and call upon him to join you in joyous anthems, were inhuman mockery and sacrilegious irony. Do you mean, citizens, to mock me, by asking me to speak today?

If so, there is a parallel to your conduct. And let me warn you that it is dangerous to copy the example of a nation whose crimes, towering up to heaven, were thrown down by the breath of the Almighty, burying that nation in irrevocable ruin! I can today take up the plaintive lament of a peeled and woe-smitten people!

By the rivers of Babylon, there we sat down. Yea! we wept when we remembered Zion. We hanged our harps upon the willows in the midst thereof. For there, they that carried us away captive required of us a song; and they who wasted us required of us mirth, saying, "Sing us one of the songs of Zion." How can we sing the Lord's song in a strange land? If I forget thee, O Jerusalem, let my right hand forget her cunning. If I do not remember thee, let my tongue cleave to the roof of my mouth.

Fellow-citizens, above your national, tumultuous joy, I hear the mournful wail of millions! whose chains, heavy and grievous yesterday, are, today, rendered more intolerable by the jubilee shouts that reach them. If I do forget, if I do not faithfully remember those bleeding children of sorrow this day, "may my right hand forget her cunning, and may my tongue cleave to the roof of my mouth!" To forget them, to pass lightly over their wrongs, and to chime in with the popular theme, would be treason most scandalous and shocking, and would make me a reproach before God and the world.

Questions:

1. What is the primary purpose of Douglass's speech?
   - A) To celebrate the principles of freedom and independence
   - B) To express the sorrow and injustice experienced by slaves
   - C) To criticize the Founding Fathers for their actions
   - D) To propose new legislation for the abolition of slavery
2. In the passage, Douglass compares the Fourth of July to:
   - A) A joyous celebration for all Americans
   - B) A painful reminder of the disparity between free and enslaved people
   - C) A symbol of hope for future freedom
   - D) An irrelevant holiday for the American people
3. The phrase "mournful wail of millions" most likely refers to:
   - A) The cry of slaves who are still in bondage
   - B) The sound of a distant battle
   - C) The sadness of those who have lost their loved ones in war
   - D) The disappointment of those who failed to achieve independence
4. What is the significance of the biblical reference to Babylon in the passage?
   - A) To highlight the religious foundation of the American nation
   - B) To draw a parallel between the plight of the Israelites and the situation of African Americans
   - C) To criticize religious hypocrisy in America
   - D) To suggest that America will face destruction like Babylon
5. How does Douglass's tone contribute to the overall impact of the speech?
   - A) His tone is celebratory, encouraging all Americans to rejoice.
   - B) His tone is mournful and accusatory, which emphasizes the deep injustice faced by slaves.
   - C) His tone is indifferent, reflecting a lack of concern for the subject.
   - D) His tone is hopeful, suggesting that freedom is close at hand.

Answers and Explanations:

1. **Question:** What is the primary purpose of Douglass's speech?
   - **Correct Answer:** B) *To express the sorrow and injustice experienced by slaves*
   - **Explanation:** Douglass's speech vividly expresses the pain and injustice felt by slaves who are excluded from the freedoms celebrated on Independence Day. His tone and content clearly aim to highlight this disparity rather than celebrate the holiday.
2. **Question:** In the passage, Douglass compares the Fourth of July to:
   - **Correct Answer:** B) *A painful reminder of the disparity between free and enslaved people*
   - **Explanation:** Douglass explicitly states that the Fourth of July is "yours, not mine," indicating that the holiday serves as a painful reminder of the freedoms that slaves do not share.
3. **Question:** The phrase "mournful wail of millions" most likely refers to:
   - **Correct Answer:** A) *The cry of slaves who are still in bondage*
   - **Explanation:** This phrase refers to the millions of enslaved people whose suffering is heightened by the celebration of a freedom that they do not enjoy. Douglass uses this imagery to underscore the deep pain and injustice experienced by slaves.
4. **Question:** What is the significance of the biblical reference to Babylon in the passage?

    - **Correct Answer:** B) *To draw a parallel between the plight of the Israelites and the situation of African Americans*
    - **Explanation:** Douglass draws a parallel between the Israelites' captivity in Babylon and the enslavement of African Americans. This comparison emphasizes the deep suffering and longing for freedom experienced by slaves, similar to that of the Israelites in exile.
5. **Question:** How does Douglass's tone contribute to the overall impact of the speech?
    - **Correct Answer:** B) *His tone is mournful and accusatory, which emphasizes the deep injustice faced by slaves.*
    - **Explanation:** Douglass's tone is somber and critical, reflecting the sorrow and anger he feels towards the celebration of a freedom that excludes millions of enslaved people. This tone intensifies the impact of his message, compelling his audience to confront the harsh realities of slavery.

Reading Test - Passage 3: Science

This passage is adapted from an article published in a scientific journal that discusses recent advancements in renewable energy technology, particularly in the field of solar power.

In recent years, the development of renewable energy sources has accelerated, driven by the urgent need to reduce carbon emissions and combat climate change. Among these sources, solar power has emerged as one of the most promising. Advances in photovoltaic (PV) cell technology have significantly increased the efficiency of solar panels, making them more cost-effective and accessible than ever before. Moreover, the integration of energy storage systems has addressed one of the primary challenges associated with solar power: its intermittent nature. Solar energy is only available when the sun is shining, which makes it unreliable without proper storage solutions. One of the most significant breakthroughs in recent years is the development of perovskite solar cells. Perovskites, a class of materials that have a unique crystal structure, have shown great potential in improving the efficiency of solar panels. Traditional silicon-based solar cells have a maximum theoretical efficiency limit of about 29%, known as the Shockley-Queisser limit. However, perovskite cells have demonstrated the potential to surpass this limit, with some laboratory prototypes achieving efficiencies of over 30%. Additionally, perovskite cells can be manufactured using simpler and cheaper processes compared to silicon cells, making them a cost-effective alternative for large-scale solar energy production.

Another promising area of research is the development of tandem solar cells, which combine two different types of PV materials to capture a broader range of the solar spectrum. By stacking a perovskite cell on top of a silicon cell, researchers have been able to create tandem cells that achieve higher efficiencies than either material alone. These tandem cells have the potential to reach efficiencies of up to 40%, making them a highly attractive option for the future of solar energy.

Energy storage technology has also seen significant advancements, particularly in the development of lithium-ion batteries. These batteries are now capable of storing large amounts of energy at a relatively low cost, making them an ideal solution for pairing with solar power systems. In addition to lithium-ion technology, researchers are exploring alternative storage methods, such as flow batteries and hydrogen fuel cells, which offer the potential for even greater energy storage capacity and longer discharge times.

The combination of these advancements in solar cell technology and energy storage systems is paving the way for a future where solar power could become the dominant global energy source. As the cost of solar energy continues to decrease and efficiency improves, it is likely that we will see a rapid expansion of solar power adoption in the coming decades. This shift has the potential to significantly reduce global carbon emissions and help mitigate the impacts of climate change.

Questions:

1. What is the main purpose of the passage?
   - A) To argue that solar power is the only viable renewable energy source
   - B) To discuss recent technological advancements that improve the efficiency and storage of solar energy
   - C) To explain the limitations of solar power compared to other energy sources
   - D) To advocate for government subsidies for solar energy research
2. According to the passage, what is one advantage of perovskite solar cells over traditional silicon-based solar cells?
   - A) They are less expensive to produce.
   - B) They are more environmentally friendly.
   - C) They have a longer lifespan.
   - D) They require less maintenance.
3. The term "intermittent nature" in the passage most likely refers to:
   - A) The fluctuating demand for solar energy.
   - B) The periodic availability of sunlight for energy generation.
   - C) The inconsistent quality of solar panels.
   - D) The variable costs associated with solar energy production.
4. What is one of the primary benefits of tandem solar cells mentioned in the passage?
   - A) They are easier to install than traditional solar cells.
   - B) They can capture a broader range of the solar spectrum.
   - C) They are less susceptible to weather-related damage.
   - D) They are more widely available than other types of solar cells.
5. The passage suggests that advancements in energy storage technology are important because they:
   - A) Make solar power more reliable by storing energy for use when the sun is not shining.
   - B) Reduce the cost of solar panel installation.
   - C) Increase the lifespan of solar panels.
   - D) Decrease the environmental impact of solar power systems.

Answers and Explanations:

1. **Question:** What is the main purpose of the passage?
   - **Correct Answer:** B) *To discuss recent technological advancements that improve the efficiency and storage of solar energy*
   - **Explanation:** The passage focuses on discussing advancements in photovoltaic technology, particularly perovskite solar cells, tandem solar cells, and energy storage systems, all of which contribute to improving the efficiency and reliability of solar energy. The passage does not advocate for solar power as the only renewable source, nor does it discuss government subsidies or directly compare solar to other energy sources.
2. **Question:** According to the passage, what is one advantage of perovskite solar cells over traditional silicon-based solar cells?
   - **Correct Answer:** A) *They are less expensive to produce.*
   - **Explanation:** The passage states that perovskite cells can be manufactured using simpler and cheaper processes compared to traditional silicon cells, making them a cost-effective alternative. This is an advantage that the passage highlights as significant for the future of solar energy.
3. **Question:** The term "intermittent nature" in the passage most likely refers to:
   - **Correct Answer:** B) *The periodic availability of sunlight for energy generation.*

- **Explanation:** The term "intermittent nature" refers to the fact that solar energy is only available when the sun is shining, which is not constant. This intermittency is a challenge for solar power, and the passage discusses how energy storage systems help address this issue.

4. **Question:** What is one of the primary benefits of tandem solar cells mentioned in the passage?
   - **Correct Answer:** B) *They can capture a broader range of the solar spectrum.*
   - **Explanation:** The passage explains that tandem solar cells combine two different types of photovoltaic materials to capture a broader range of the solar spectrum, leading to higher efficiency. This ability to capture more of the sun's energy is the primary benefit discussed.
5. **Question:** The passage suggests that advancements in energy storage technology are important because they:
   - **Correct Answer:** A) *Make solar power more reliable by storing energy for use when the sun is not shining.*
   - **Explanation:** The passage emphasizes the role of energy storage technology in addressing the intermittent nature of solar power. By storing energy for use when the sun isn't shining, these advancements make solar power a more reliable energy source.

Reading Test - Passage 4: Paired Passages (History/Social Studies)

These paired passages are adapted from two historical speeches. Passage 1 is an excerpt from "The Gettysburg Address" by President Abraham Lincoln, delivered on November 19, 1863. Passage 2 is from "A House Divided," a speech given by Abraham Lincoln on June 16, 1858, upon accepting the Illinois Republican Party's nomination for U.S. Senator.

Passage 1

Fourscore and seven years ago our fathers brought forth on this continent, a new nation, conceived in Liberty, and dedicated to the proposition that all men are created equal.

Now we are engaged in a great civil war, testing whether that nation, or any nation so conceived and so dedicated, can long endure. We are met on a great battlefield of that war. We have come to dedicate a portion of that field, as a final resting place for those who here gave their lives that that nation might live. It is altogether fitting and proper that we should do this.

But, in a larger sense, we can not dedicate—we can not consecrate—we can not hallow—this ground. The brave men, living and dead, who struggled here, have consecrated it, far above our poor power to add or detract. The world will little note, nor long remember what we say here, but it can never forget what they did here. It is for us the living, rather, to be dedicated here to the unfinished work which they who fought here have thus far so nobly advanced. It is rather for us to be here dedicated to the great task remaining before us—that from these honored dead we take increased devotion to that cause for which they gave the last full measure of devotion—that we here highly resolve that these dead shall not have died in vain—that this nation, under God, shall have a new birth of freedom—and that government of the people, by the people, for the people, shall not perish from the earth.

Passage 2

If we could first know where we are, and whither we are tending, we could then better judge what to do, and how to do it. We are now far into the fifth year since a policy was initiated with the avowed object, and confident promise, of putting an end to slavery agitation. Under the operation of that policy, that agitation has not only not ceased, but has constantly augmented. In my opinion, it will not cease until a crisis shall have been reached and passed.

"A house divided against itself cannot stand." I believe this government cannot endure, permanently, half slave and half free. I do not expect the Union to be dissolved—I do not expect the house to fall—but I do expect it will cease to be divided. It will become all one thing, or all the other. Either the opponents of slavery will arrest the further spread of it, and place it where the public mind shall rest in the belief that it is in the course of ultimate extinction; or its advocates will push it forward, till it shall become alike lawful in all the States, old as well as new—North as well as South.

Have we no tendency to the latter condition?

Questions:

1. **Both Passage 1 and Passage 2 address the theme of national unity. How do the speakers' approaches to this theme differ?**
   - A) Passage 1 is more focused on honoring the past, while Passage 2 is more concerned with predicting the future.
   - B) Passage 1 calls for immediate action, while Passage 2 reflects on the accomplishments of the past.
   - C) Passage 1 emphasizes the inevitability of conflict, while Passage 2 advocates for peace.
   - D) Passage 1 focuses on the importance of freedom, while Passage 2 downplays the issue of slavery.
2. **Which of the following best describes the tone of Passage 1?**
   - A) Humble and reverent
   - B) Urgent and demanding
   - C) Celebratory and joyful
   - D) Defiant and challenging
3. **In Passage 2, what is the significance of the phrase "A house divided against itself cannot stand"?**
   - A) It suggests that internal conflict will lead to the collapse of the nation.
   - B) It implies that slavery will continue to be accepted throughout the Union.
   - C) It argues that peace between the North and South is achievable.
   - D) It warns against the consequences of foreign intervention in American affairs.
4. **In both passages, what is the central concern regarding the future of the nation?**
   - A) The need for economic reform
   - B) The preservation of the Union
   - C) The threat of foreign invasion
   - D) The expansion of territorial boundaries
5. **Which of the following statements best summarizes Lincoln's vision for the United States as expressed in both passages?**
   - A) The nation must focus on economic prosperity to ensure its future.
   - B) The United States must remain unified and free from slavery to endure.
   - C) The nation should expand its influence globally to maintain its power.
   - D) The government should prioritize maintaining peace at all costs.

Answers and Explanations:

1. **Question:** Both Passage 1 and Passage 2 address the theme of national unity. How do the speakers' approaches to this theme differ?
    - **Correct Answer:** A) *Passage 1 is more focused on honoring the past, while Passage 2 is more concerned with predicting the future.*
    - **Explanation:** In Passage 1, Lincoln focuses on honoring the sacrifices made during the Civil War and dedicating the living to the ongoing struggle for freedom. Passage 2, on the other hand, is more forward-looking, with Lincoln predicting the future outcome of the nation's division over slavery and calling attention to the inevitable crisis that must be resolved.
2. **Question:** Which of the following best describes the tone of Passage 1?
    - **Correct Answer:** A) *Humble and reverent*
    - **Explanation:** The tone of "The Gettysburg Address" is humble and reverent as Lincoln honors the soldiers who fought at Gettysburg and expresses deep respect for their sacrifice. He acknowledges that their actions have consecrated the ground far beyond what words could do.
3. **Question:** In Passage 2, what is the significance of the phrase "A house divided against itself cannot stand"?
    - **Correct Answer:** A) *It suggests that internal conflict will lead to the collapse of the nation.*
    - **Explanation:** Lincoln uses the metaphor of a divided house to convey that the nation cannot endure permanently half free and half slave. He argues that the Union must become all one thing or all the other—either entirely free or entirely enslaved—indicating that the current division is unsustainable and will eventually lead to a crisis.
4. **Question:** In both passages, what is the central concern regarding the future of the nation?
    - **Correct Answer:** B) *The preservation of the Union*
    - **Explanation:** Both passages are deeply concerned with the survival of the United States as a unified nation. In Passage 1, Lincoln emphasizes the importance of ensuring that the nation endures through dedication to the cause of freedom. In Passage 2, Lincoln warns that the nation cannot survive in its divided state and must resolve the issue of slavery to preserve the Union.
5. **Question:** Which of the following statements best summarizes Lincoln's vision for the United States as expressed in both passages?
    - **Correct Answer:** B) *The United States must remain unified and free from slavery to endure.*
    - **Explanation:** Lincoln's vision for the nation, as expressed in both speeches, is that the Union must be preserved and that the nation should be committed to the principles of freedom and equality. He believes that the nation cannot survive if it remains divided over the issue of slavery, and both passages reflect his commitment to these ideals.

Reading Test - Passage 5: Science

This passage is adapted from an article discussing the discovery of a new exoplanet in a distant star system and its implications for the search for extraterrestrial life.

In recent years, the discovery of exoplanets—planets that orbit stars outside our solar system—has revolutionized our understanding of the universe. One of the most exciting developments in this field occurred when a team of astronomers discovered an Earth-like exoplanet, designated Kepler-452b, located in the habitable zone of its parent star, Kepler-452. The habitable zone, often referred to as the "Goldilocks zone," is the region around a star where conditions are just right for liquid water to exist—a crucial ingredient for life as we know it.

Kepler-452b, often dubbed "Earth's cousin," is about 60% larger in diameter than Earth and has a mass approximately five times that of our planet. Its orbit around Kepler-452 is remarkably similar to Earth's orbit around the Sun, taking about 385 days to complete a single revolution. The star Kepler-452 is slightly older and more luminous than our Sun, suggesting that Kepler-452b may have had a longer period of stable conditions suitable for life.

While the discovery of Kepler-452b is a significant milestone, it also raises new questions and challenges. One of the primary concerns is the planet's distance from Earth—approximately 1,400 light-years away. This vast distance makes it impossible to directly observe the planet in detail with current technology. Instead, scientists rely on indirect methods, such as measuring the light dimming as the planet passes in front of its star (a method known as the transit method), to infer its properties.

Another challenge is determining whether Kepler-452b has an atmosphere and, if so, what its composition might be. The presence of a stable atmosphere is essential for life as we understand it, as it can regulate temperature, protect against harmful radiation, and potentially support complex chemical processes. Future missions, such as the James Webb Space Telescope, are expected to provide more detailed observations that could help answer these questions.

The discovery of Kepler-452b has fueled speculation about the possibility of life beyond our solar system. While the existence of life on Kepler-452b remains purely speculative, the fact that such a planet exists in the habitable zone of a Sun-like star provides hope that other Earth-like worlds may be out there, waiting to be discovered. The ongoing search for exoplanets continues to push the boundaries of our knowledge, bringing us closer to answering one of humanity's oldest questions: Are we alone in the universe?

Questions:

1. What is the main focus of the passage?

- A) The methods used to discover exoplanets
- B) The challenges of observing distant star systems
- C) The discovery and implications of an Earth-like exoplanet
- D) The technological advancements in space exploration

2. The term "Goldilocks zone" in the passage refers to:
   - A) A region of a planet where temperatures are extremely high
   - B) The area around a star where conditions are suitable for liquid water
   - C) A method used to measure the distance of a planet from its star
   - D) The time period when a star is most luminous
3. Why is Kepler-452b referred to as "Earth's cousin" in the passage?
   - A) It has a similar size and mass to Earth.
   - B) It has an atmosphere similar to Earth's.
   - C) It orbits a star in a manner similar to Earth's orbit around the Sun.
   - D) It is located within our solar system.
4. According to the passage, why is the distance of Kepler-452b from Earth a significant challenge for scientists?
   - A) It prevents the use of the transit method to study the planet.
   - B) It limits the ability to send spacecraft to explore the planet.
   - C) It causes interference with telescopic observations.
   - D) It results in the planet being outside the habitable zone.
5. What is the significance of the discovery of Kepler-452b, as suggested by the passage?
   - A) It confirms the presence of life on other planets.
   - B) It provides evidence that Earth-like planets may be common in the universe.
   - C) It demonstrates that all planets in the habitable zone can support life.
   - D) It shows that stars similar to the Sun are rare in the universe.

Answers and Explanations:

1. **Question:** What is the main focus of the passage?
   - **Correct Answer:** C) *The discovery and implications of an Earth-like exoplanet*
   - **Explanation:** The passage primarily discusses the discovery of Kepler-452b, an exoplanet in the habitable zone of its star, and explores the potential implications of this discovery for the search for extraterrestrial life.
2. **Question:** The term "Goldilocks zone" in the passage refers to:
   - **Correct Answer:** B) *The area around a star where conditions are suitable for liquid water*
   - **Explanation:** The "Goldilocks zone" is the region around a star where temperatures are "just right" for liquid water to exist, which is considered essential for life as we know it. The passage explains that Kepler-452b is located in this zone.
3. **Question:** Why is Kepler-452b referred to as "Earth's cousin" in the passage?
   - **Correct Answer:** C) *It orbits a star in a manner similar to Earth's orbit around the Sun.*
   - **Explanation:** Kepler-452b is called "Earth's cousin" because it has an orbit around its star that is similar to Earth's orbit around the Sun, making it an Earth-like exoplanet in terms of its potential to support life.
4. **Question:** According to the passage, why is the distance of Kepler-452b from Earth a significant challenge for scientists?
   - **Correct Answer:** B) *It limits the ability to send spacecraft to explore the planet.*
   - **Explanation:** The passage explains that Kepler-452b is approximately 1,400 light-years away, making it impossible with current technology to send spacecraft to explore the planet or observe it directly in detail. This vast distance poses a significant challenge for scientists.

5. **Question:** What is the significance of the discovery of Kepler-452b, as suggested by the passage?
    - **Correct Answer:** B) *It provides evidence that Earth-like planets may be common in the universe.*
    - **Explanation:** The discovery of Kepler-452b, an Earth-like planet in the habitable zone of a Sun-like star, suggests that such planets may be more common than previously thought. This discovery fuels hope that there may be other Earth-like worlds capable of supporting life.

Writing and Language Test

Passage 1: Careers

This passage is adapted from an article discussing the benefits of remote work and its impact on employee productivity and job satisfaction.

[1] The shift to remote work has been one of the most significant changes in the modern workplace. **[2]** With advancements in technology, many companies have adopted remote work policies, allowing employees to work from home or other locations outside of the traditional office. **[3]** This change has not only improved work-life balance for employees but also increased productivity and job satisfaction. **[4]** However, remote work also presents challenges, such as feelings of isolation and difficulties in collaboration.

[5] Research indicates that employees who work remotely are often more productive than their office-bound counterparts. **[6]** They save time by not commuting, which they can spend on work or personal activities. **[7]** Additionally, remote workers often have more flexibility in managing their schedules, which can lead to increased efficiency and job satisfaction.

[8] Despite these benefits, remote work is not without its drawbacks. **[9]** Some employees report feeling disconnected from their colleagues, leading to feelings of isolation. **[10]** Moreover, virtual communication tools, while useful, may not fully replicate the experience of face-to-face interactions, which can hinder effective collaboration.

[11] In conclusion, while remote work offers numerous benefits, it is essential for companies to address the challenges it presents. **[12]** By providing adequate support and resources, such as regular check-ins and team-building activities, employers can help remote workers stay connected and engaged. **[13]** As the trend towards remote work continues, companies must find ways to balance the advantages with the challenges to create a productive and satisfying work environment for all employees.

Questions:

1. **Which choice best maintains the style and tone of the passage in the first sentence of paragraph 1?**
    - A) The switch to remote work is, like, the biggest change ever in the workplace.

- B) The transition to remote work has been one of the most significant developments in the contemporary workplace.
- C) Remote work is becoming a popular option for employees everywhere.
- D) The move to remote work is a pretty big deal in today's workplace.

2. **In sentence 4, the writer wants to acknowledge the potential downsides of remote work. Which choice most effectively accomplishes this goal?**
   - A) However, remote work can be boring for some employees.
   - B) However, remote work isn't always as fun as it seems.
   - C) However, remote work also presents challenges, such as feelings of isolation and difficulties in collaboration.
   - D) However, not everyone likes working remotely.
3. **In sentence 6, which choice provides the most specific and accurate information?**
   - A) They save time by not commuting, which they can spend on work or personal activities.
   - B) They have more free time because they don't have to travel to work.
   - C) Skipping the commute gives them lots of extra time.
   - D) Not commuting means they have more time for stuff they enjoy.
4. **Which choice best supports the conclusion drawn in sentence 13?**
   - A) Companies must understand that remote work is a growing trend.
   - B) The best approach is to find a balance between remote work and office work.
   - C) Employers should consider the needs of their employees when implementing remote work policies.
   - D) As the trend towards remote work continues, companies must find ways to balance the advantages with the challenges to create a productive and satisfying work environment for all employees.
5. **Which choice most effectively combines sentences 7 and 8?**
   - A) Remote workers often have more flexibility in managing their schedules, but there are also drawbacks to working remotely.
   - B) While remote workers often have more flexibility in managing their schedules, the arrangement is not without its drawbacks.
   - C) Remote work offers flexibility in managing schedules and has many drawbacks.
   - D) Although remote work provides flexibility, it also presents significant challenges that need to be addressed.

Answers and Explanations:

1. **Question:** Which choice best maintains the style and tone of the passage in the first sentence of paragraph 1?
   - **Correct Answer:** B) *The transition to remote work has been one of the most significant developments in the contemporary workplace.*
   - **Explanation:** Choice B uses formal and precise language that is consistent with the style and tone of the passage. The other options are either too informal or lack the appropriate tone for the context.
2. **Question:** In sentence 4, the writer wants to acknowledge the potential downsides of remote work. Which choice most effectively accomplishes this goal?
   - **Correct Answer:** C) *However, remote work also presents challenges, such as feelings of isolation and difficulties in collaboration.*
   - **Explanation:** Choice C clearly and specifically addresses the potential challenges of remote work, such as isolation and collaboration issues, which aligns with the purpose of acknowledging the downsides.

3. **Question:** In sentence 6, which choice provides the most specific and accurate information?
   - **Correct Answer:** A) *They save time by not commuting, which they can spend on work or personal activities.*
   - **Explanation:** Choice A provides a clear, specific, and accurate explanation of how remote workers benefit from the time saved by not commuting. The other choices are either vague or lack the precision needed to support the point effectively.
4. **Question:** Which choice best supports the conclusion drawn in sentence 13?
   - **Correct Answer:** D) *As the trend towards remote work continues, companies must find ways to balance the advantages with the challenges to create a productive and satisfying work environment for all employees.*
   - **Explanation:** Choice D directly ties into the conclusion by emphasizing the need to balance the benefits and challenges of remote work, which is the main idea of the conclusion. It effectively summarizes the key points discussed in the passage.
5. **Question:** Which choice most effectively combines sentences 7 and 8?
   - **Correct Answer:** B) *While remote workers often have more flexibility in managing their schedules, the arrangement is not without its drawbacks.*
   - **Explanation:** Choice B effectively combines the two sentences by acknowledging the flexibility remote work offers while also introducing the idea that there are challenges. This combination maintains the logical flow and coherence of the passage.

Passage 2: Humanities

This passage is adapted from an article discussing the importance of preserving cultural heritage sites around the world.

[1] Cultural heritage sites, such as ancient ruins, historic buildings, and sacred landscapes, are invaluable resources that connect us to our past. **[2]** They provide insights into the history, beliefs, and traditions of different cultures, and they serve as a source of pride and identity for communities around the world. **[3]** However, these sites are increasingly threatened by various factors, including urbanization, climate change, and tourism.

[4] Urbanization often leads to the destruction of heritage sites, as modern infrastructure projects take precedence over preservation efforts. **[5]** For instance, the construction of highways, shopping centers, and residential areas frequently encroaches upon or even completely obliterates historic landmarks. **[6]** Climate change poses another significant threat, as rising temperatures, increased rainfall, and more frequent natural disasters can cause irreparable damage to fragile structures and landscapes.

[7] Tourism, although economically beneficial, can also have negative effects on cultural heritage sites. **[8]** The influx of visitors often leads to overcrowding, which can result in physical wear and tear on structures that were never designed to accommodate large numbers of people. **[9]** Additionally, the commercialization of cultural sites can lead to the erosion of their historical and cultural significance, as they are transformed into mere attractions for tourists.

[10] To protect these irreplaceable sites, it is crucial to implement strategies that balance preservation with development and tourism. **[11]** This may include enforcing stricter regulations on construction near heritage sites, investing in conservation and restoration projects, and promoting sustainable tourism practices that minimize the impact on these sites. **[12]** Public awareness campaigns can also play a vital role in educating people about the importance of preserving cultural heritage and encouraging them to support conservation efforts.

Questions:

1. **In the context of the passage, which choice best introduces the topic in sentence 1?**
   - A) Cultural heritage sites can be found all over the world, from ancient ruins to modern monuments.
   - B) Cultural heritage sites are crucial to understanding the history and traditions of various societies.
   - C) Many people enjoy visiting cultural heritage sites during their vacations.
   - D) Cultural heritage sites are old and interesting places.
2. **In sentence 4, the author wants to emphasize the impact of urbanization on heritage sites. Which choice most effectively achieves this goal?**
   - A) Urbanization often leads to the destruction of heritage sites, as modern infrastructure projects take precedence over preservation efforts.

- B) Urbanization sometimes causes changes to heritage sites, which may or may not be beneficial.
- C) Urbanization is a growing trend that can influence the preservation of cultural heritage sites.
- D) Urbanization affects many aspects of modern life, including the preservation of cultural heritage sites.

3. **Which choice most effectively combines sentences 5 and 6?**
   - A) Modern infrastructure projects like highways, shopping centers, and residential areas frequently encroach upon or completely obliterate historic landmarks; climate change, with its rising temperatures, increased rainfall, and more frequent natural disasters, also poses a significant threat.
   - B) Both modern infrastructure projects, which frequently encroach upon or completely obliterate historic landmarks, and climate change, which causes rising temperatures, increased rainfall, and more frequent natural disasters, pose significant threats to cultural heritage sites.
   - C) The construction of highways, shopping centers, and residential areas frequently encroaches upon or obliterates historic landmarks, while climate change, with its rising temperatures, increased rainfall, and more frequent natural disasters, poses another significant threat.
   - D) The construction of highways, shopping centers, and residential areas frequently encroaches upon or even obliterates historic landmarks; climate change, with its rising temperatures, increased rainfall, and more frequent natural disasters, poses another significant threat.
4. **In sentence 9, which choice most effectively supports the argument that tourism can negatively affect cultural heritage sites?**
   - A) Additionally, the commercialization of cultural sites can lead to the erosion of their historical and cultural significance, as they are transformed into mere attractions for tourists.
   - B) Additionally, tourism can change the way people view cultural heritage sites, sometimes in a negative way.
   - C) Additionally, tourism can be both beneficial and harmful to cultural heritage sites, depending on how it is managed.
   - D) Additionally, the popularity of cultural heritage sites often leads to a loss of authenticity.
5. **Which choice most effectively concludes the passage?**
   - A) Public awareness campaigns can also play a vital role in educating people about the importance of preserving cultural heritage and encouraging them to support conservation efforts.
   - B) Ultimately, the best way to protect cultural heritage sites is to ensure that they are carefully maintained and monitored.
   - C) The preservation of cultural heritage sites is important for future generations, who deserve to learn about and appreciate these valuable resources.
   - D) If we fail to protect cultural heritage sites, we risk losing an essential part of our history and identity.

Answers and Explanations:

1. **Question:** In the context of the passage, which choice best introduces the topic in sentence 1?
   - **Correct Answer:** B) *Cultural heritage sites are crucial to understanding the history and traditions of various societies.*
   - **Explanation:** Choice B best introduces the passage by highlighting the significance of cultural heritage sites in understanding history and traditions, which aligns with the main idea of the passage. The other choices are either too general or not directly related to the main focus of the passage.
2. **Question:** In sentence 4, the author wants to emphasize the impact of urbanization on heritage sites. Which choice most effectively achieves this goal?
   - **Correct Answer:** A) *Urbanization often leads to the destruction of heritage sites, as modern infrastructure projects take precedence over preservation efforts.*

- **Explanation:** Choice A clearly emphasizes the negative impact of urbanization on heritage sites by stating that modern infrastructure projects often lead to their destruction. This choice aligns with the passage's emphasis on the threats posed by urbanization.

3. **Question:** Which choice most effectively combines sentences 5 and 6?
   - **Correct Answer:** C) *The construction of highways, shopping centers, and residential areas frequently encroaches upon or obliterates historic landmarks, while climate change, with its rising temperatures, increased rainfall, and more frequent natural disasters, poses another significant threat.*
   - **Explanation:** Choice C effectively combines the two sentences by clearly linking the threats posed by both urbanization and climate change. The structure is logical and maintains the focus on the significant challenges to heritage sites.
4. **Question:** In sentence 9, which choice most effectively supports the argument that tourism can negatively affect cultural heritage sites?
   - **Correct Answer:** A) *Additionally, the commercialization of cultural sites can lead to the erosion of their historical and cultural significance, as they are transformed into mere attractions for tourists.*
   - **Explanation:** Choice A provides a strong argument by explaining how tourism can erode the historical and cultural significance of sites when they are commercialized and reduced to mere tourist attractions. This effectively supports the idea that tourism can have negative effects.
5. **Question:** Which choice most effectively concludes the passage?
   - **Correct Answer:** D) *If we fail to protect cultural heritage sites, we risk losing an essential part of our history and identity.*
   - **Explanation:** Choice D effectively concludes the passage by emphasizing the high stakes involved in preserving cultural heritage sites. It highlights the potential loss of history and identity, which reinforces the passage's main argument about the importance of preservation.

Passage 3: Science

This passage is adapted from an article discussing recent advances in biotechnology, specifically the development of CRISPR technology and its potential applications.

[1] CRISPR technology, which stands for "Clustered Regularly Interspaced Short Palindromic Repeats," has revolutionized the field of genetics. **[2]** It allows scientists to precisely edit DNA, making it possible to add, remove, or alter genetic material at specific locations in the genome. **[3]** Since its discovery, CRISPR has been hailed as a groundbreaking tool with numerous potential applications, from curing genetic diseases to enhancing agricultural crops.

[4] One of the most promising applications of CRISPR is in the field of medicine. **[5]** Scientists are exploring the possibility of using CRISPR to treat genetic disorders such as cystic fibrosis, muscular dystrophy, and sickle cell anemia. **[6]** By targeting the specific genes responsible for these conditions, researchers hope to correct the underlying genetic mutations and potentially cure these diseases. **[7]** While these developments are still in the experimental stage, early results are encouraging.

[8] Another exciting application of CRISPR is in agriculture. **[9]** By modifying the genes of crops, scientists can create plants that are more resistant to pests, diseases, and environmental stresses. **[10]** This could lead to higher crop yields and more sustainable farming practices, particularly in regions where food security is a pressing concern.

[11] However, the use of CRISPR technology raises important ethical questions. **[12]** Some critics argue that editing the human genome could have unforeseen consequences, leading to unintended side effects or even new genetic disorders. **[13]** There are also concerns about the potential for CRISPR to be used for non-therapeutic purposes, such as "designer babies," where genetic traits are selected based on personal preferences.

[14] Despite these concerns, the potential benefits of CRISPR technology are immense. **[15]** With proper regulation and oversight, it is possible to harness the power of CRISPR for the greater good, ensuring that its applications are safe, ethical, and beneficial for society as a whole.

Questions:

1. **Which choice best maintains the tone and style of the passage in sentence 1?**
    - A) CRISPR technology is like, really cool because it lets scientists mess around with DNA.
    - B) CRISPR technology is a fascinating innovation that has significantly impacted genetics.
    - C) CRISPR technology, which stands for a bunch of complicated scientific terms, is pretty amazing.
    - D) CRISPR technology is a new tool that lets scientists edit DNA.

2. **In sentence 6, which choice most effectively explains the purpose of using CRISPR in treating genetic disorders?**
   - A) By targeting the specific genes responsible for these conditions, researchers hope to correct the underlying genetic mutations and potentially cure these diseases.
   - B) By using CRISPR, scientists can look at genes and try to figure out what's wrong with them.
   - C) By using CRISPR, researchers can remove harmful genes and replace them with good ones.
   - D) By targeting DNA, researchers hope to cure genetic diseases.
3. **In sentence 9, which choice provides the most specific and accurate information?**
   - A) By modifying the genes of crops, scientists can make plants that are better at handling tough conditions.
   - B) By modifying the genes of crops, scientists can create plants that are more resistant to pests, diseases, and environmental stresses.
   - C) By modifying crops, scientists can make them grow bigger and faster.
   - D) By using CRISPR, scientists can make crops that are more awesome.
4. **Which choice most effectively combines sentences 7 and 8?**
   - A) Early results are encouraging, but these developments are still in the experimental stage, and CRISPR also has exciting applications in agriculture.
   - B) While these developments are still in the experimental stage, early results are encouraging; CRISPR is also being used in agriculture to improve crop resilience.
   - C) Early results are encouraging, but CRISPR is also used in agriculture, where scientists are working to improve crop yields.
   - D) While CRISPR is still being tested in medicine, it's already being used in agriculture to make crops better.
5. **Which choice most effectively introduces the ethical concerns discussed in sentence 11?**
   - A) Some people think CRISPR is great, but others worry about its impact.
   - B) Not everyone is excited about CRISPR; there are some big ethical issues to consider.
   - C) The use of CRISPR in medicine and agriculture has sparked important ethical debates.
   - D) CRISPR is cool, but it also has some downsides.

Answers and Explanations:

1. **Question:** Which choice best maintains the tone and style of the passage in sentence 1?
   - **Correct Answer:** B) *CRISPR technology is a fascinating innovation that has significantly impacted genetics.*
   - **Explanation:** Choice B uses formal and precise language that is consistent with the academic tone of the passage. It accurately reflects the impact of CRISPR on the field of genetics. The other choices are either too informal or do not convey the significance of CRISPR as effectively.
2. **Question:** In sentence 6, which choice most effectively explains the purpose of using CRISPR in treating genetic disorders?
   - **Correct Answer:** A) *By targeting the specific genes responsible for these conditions, researchers hope to correct the underlying genetic mutations and potentially cure these diseases.*
   - **Explanation:** Choice A clearly explains the goal of using CRISPR to target and correct specific genetic mutations, which aligns with the passage's focus on the potential for curing genetic disorders. It provides a detailed and accurate explanation compared to the other options.
3. **Question:** In sentence 9, which choice provides the most specific and accurate information?
   - **Correct Answer:** B) *By modifying the genes of crops, scientists can create plants that are more resistant to pests, diseases, and environmental stresses.*
   - **Explanation:** Choice B provides specific and relevant details about how CRISPR can be used in agriculture, focusing on enhancing the resilience of crops to various challenges. This choice is more specific and accurate than the other options.

4. **Question:** Which choice most effectively combines sentences 7 and 8?
   - **Correct Answer:** B) *While these developments are still in the experimental stage, early results are encouraging; CRISPR is also being used in agriculture to improve crop resilience.*
   - **Explanation:** Choice B effectively combines the sentences by maintaining the logical flow between the experimental stage of CRISPR in medicine and its applications in agriculture. This combination is clear and concise, preserving the connection between the two ideas.
5. **Question:** Which choice most effectively introduces the ethical concerns discussed in sentence 11?
   - **Correct Answer:** C) *The use of CRISPR in medicine and agriculture has sparked important ethical debates.*
   - **Explanation:** Choice C introduces the ethical concerns in a formal and precise manner, setting the stage for the discussion of potential risks and controversies surrounding CRISPR technology. This choice is the most appropriate for maintaining the academic tone of the passage.

Passage 4: History/Social Studies

This passage is adapted from an article discussing the impact of the Industrial Revolution on society and the economy.

[1] The Industrial Revolution, which began in the late 18th century, marked a significant turning point in history. **[2]** It transformed economies that had been based on agriculture and handicrafts into economies based on large-scale industry, mechanized manufacturing, and the factory system. **[3]** The revolution led to the rise of urbanization, as people moved from rural areas to cities in search of work in the new factories. **[4]** It also had profound effects on the social structure, labor, and the global economy.

[5] One of the most significant changes brought about by the Industrial Revolution was the shift from agrarian societies to industrialized ones. **[6]** Before the revolution, most people lived in rural areas and worked in agriculture or small-scale crafts. **[7]** However, with the advent of new machinery and production techniques, large factories became the primary means of production. **[8]** This shift led to a massive migration of people to urban centers, where they could find employment in the growing number of factories.

[9] The Industrial Revolution also had a significant impact on the labor force. **[10]** The factory system introduced a new way of working, with workers performing specialized tasks on a production line. **[11]** This was a stark contrast to the artisanal model, where craftspeople were involved in the entire process of making a product. **[12]** The division of labor in factories led to increased productivity but also to harsher working conditions, as workers were often subjected to long hours, low wages, and unsafe environments.

[13] Another important aspect of the Industrial Revolution was its effect on the global economy. **[14]** The mass production of goods led to a decrease in prices, making products more accessible to a broader population. **[15]** This, in turn, fueled consumer demand and further stimulated industrial growth. **[16]** Additionally, the expansion of trade networks and the development of new markets allowed industrialized nations to increase their wealth and influence on the world stage.

[17] However, the benefits of the Industrial Revolution were not equally distributed. **[18]** While some people prospered, others faced significant hardships. **[19]** The rapid urbanization and industrialization of society led to overcrowded cities, poor living conditions, and environmental pollution. **[20]** Moreover, the exploitation of labor, including child labor, became a widespread issue, sparking social reform movements and calls for better working conditions.

Questions:

1. **Which choice best introduces the central theme of the passage in sentence 1?**

- A) The Industrial Revolution brought about many inventions and innovations.
- B) The Industrial Revolution, which began in the late 18th century, marked a significant turning point in history.
- C) The Industrial Revolution was a period of great change and advancement.
- D) The Industrial Revolution changed everything.

2. **In sentence 8, which choice most effectively emphasizes the magnitude of migration to urban centers?**
   - A) This shift led to a large migration of people to urban centers.
   - B) This shift led to some people moving to urban centers.
   - C) This shift led to a massive migration of people to urban centers.
   - D) This shift caused people to move to urban centers.
3. **Which choice best clarifies the contrast between the factory system and the artisanal model in sentence 11?**
   - A) The factory system was different from the artisanal model because it used more advanced technology.
   - B) The factory system involved large-scale production, whereas the artisanal model was based on small-scale crafts.
   - C) The factory system was about mass production, while the artisanal model focused on creating unique items.
   - D) The factory system relied on specialization, in contrast to the artisanal model, where craftspeople were involved in the entire process.
4. **Which choice most effectively combines sentences 15 and 16?**
   - A) This, in turn, fueled consumer demand and further stimulated industrial growth, as the expansion of trade networks and the development of new markets allowed industrialized nations to increase their wealth and influence on the world stage.
   - B) This, in turn, fueled consumer demand and further stimulated industrial growth; moreover, the expansion of trade networks and the development of new markets allowed industrialized nations to increase their wealth and influence on the world stage.
   - C) As a result of this, consumer demand grew and stimulated industrial growth, and the expansion of trade networks and the development of new markets also allowed industrialized nations to increase their wealth and influence on the world stage.
   - D) This fueled consumer demand and stimulated industrial growth, while the expansion of trade networks and the development of new markets allowed industrialized nations to increase their wealth and influence on the world stage.
5. **Which choice most effectively concludes the passage?**
   - A) As we reflect on the Industrial Revolution, it is important to recognize both its benefits and its drawbacks.
   - B) The Industrial Revolution brought about great changes, but it also introduced significant challenges that still resonate today.
   - C) In conclusion, the Industrial Revolution was a period of significant change and progress for society.
   - D) Ultimately, the Industrial Revolution was a time of transformation that reshaped the world in both positive and negative ways.

Answers and Explanations:

1. **Question:** Which choice best introduces the central theme of the passage in sentence 1?
   - **Correct Answer:** B) *The Industrial Revolution, which began in the late 18th century, marked a significant turning point in history.*

- **Explanation:** Choice B effectively introduces the passage by highlighting the significance of the Industrial Revolution as a major historical event. It sets the stage for discussing the widespread changes brought about by the revolution, which aligns with the passage's focus.

2. **Question:** In sentence 8, which choice most effectively emphasizes the magnitude of migration to urban centers?
    - **Correct Answer:** C) *This shift led to a massive migration of people to urban centers.*
    - **Explanation:** Choice C emphasizes the large scale of migration that occurred as people moved to cities during the Industrial Revolution. The use of the word "massive" effectively conveys the extent of this migration, which is consistent with the historical context.
3. **Question:** Which choice best clarifies the contrast between the factory system and the artisanal model in sentence 11?
    - **Correct Answer:** D) *The factory system relied on specialization, in contrast to the artisanal model, where craftspeople were involved in the entire process.*
    - **Explanation:** Choice D clearly contrasts the factory system, which involved specialized tasks, with the artisanal model, where craftspeople handled all aspects of production. This choice provides a precise comparison that aligns with the passage's discussion of the differences between the two systems.
4. **Question:** Which choice most effectively combines sentences 15 and 16?
    - **Correct Answer:** B) *This, in turn, fueled consumer demand and further stimulated industrial growth; moreover, the expansion of trade networks and the development of new markets allowed industrialized nations to increase their wealth and influence on the world stage.*
    - **Explanation:** Choice B combines the sentences in a way that maintains the logical flow of ideas and clearly connects the impact of consumer demand with the expansion of trade networks. The use of "moreover" effectively links the two related ideas, making the combination smooth and coherent.
5. **Question:** Which choice most effectively concludes the passage?
    - **Correct Answer:** D) *Ultimately, the Industrial Revolution was a time of transformation that reshaped the world in both positive and negative ways.*
    - **Explanation:** Choice D provides a balanced conclusion that acknowledges both the positive and negative effects of the Industrial Revolution, which is consistent with the passage's overall discussion of its wide-ranging impact. This choice effectively summarizes the key points and leaves the reader with a thoughtful reflection on the revolution's legacy.

Tips and Tricks for Excelling in the SAT Reading and Writing Sections

The SAT Reading and Writing sections test your ability to comprehend, analyze, and improve written material. By mastering key strategies and understanding the nuances of these sections, you can significantly boost your score. Below are essential tips and tricks to help you excel.

Reading Test Tips and Tricks

1. **Skim Passages First:**
   - Before diving into the questions, quickly skim the passage to get a sense of the main idea, tone, and structure. Focus on the first and last sentences of paragraphs, as these often contain key points.
   - **Tip:** Underline or take mental notes of key details as you skim. This will make it easier to locate information when answering questions.
2. **Focus on the Big Picture:**
   - Many questions ask about the main idea, purpose, or tone of the passage. Pay attention to the author's overall argument or narrative arc.
   - **Tip:** After reading, summarize the passage in one sentence. This will help you answer big-picture questions more accurately.
3. **Use Context Clues for Vocabulary Questions:**
   - Vocabulary in context questions ask you to determine the meaning of a word based on how it is used in the passage. Look at the sentences before and after the word to gather clues.
   - **Tip:** Substitute each answer choice into the sentence to see which one makes the most sense in the context.
4. **Answer Line-Referenced Questions First:**
   - Questions that reference specific lines in the passage can often be answered quickly. These questions provide a direct clue about where to look, making them a good starting point.
   - **Tip:** Always read a few lines before and after the referenced text to understand the full context.
5. **Identify and Focus on Keywords:**
   - Questions often include keywords that hint at what the question is really asking. Words like "mainly," "suggests," or "most likely" can guide your interpretation.
   - **Tip:** Use the process of elimination to weed out answer choices that don't directly address the keywords in the question.
6. **Don't Get Stuck on Difficult Questions:**
   - If a question seems particularly challenging, skip it and return to it later. Spending too much time on one question can hurt your overall performance.
   - **Tip:** Mark questions that you skip so you can easily return to them before time runs out.
7. **Practice Active Reading:**

- Engage with the passage as you read by asking yourself questions like, "What is the author trying to convey?" and "How does this paragraph support the main idea?"
- **Tip:** Consider the author's perspective and any potential biases. This will help you better understand the purpose and tone of the passage.

Writing and Language Test Tips and Tricks

1. **Focus on Clarity and Conciseness:**
   - The SAT Writing and Language section values clear and concise writing. Look for answer choices that eliminate redundancy and simplify complex ideas.
   - **Tip:** Avoid choices that add unnecessary words or repeat ideas. The most direct answer is usually the correct one.
2. **Understand Common Grammar Rules:**
   - Questions often test your knowledge of grammar rules such as subject-verb agreement, parallel structure, and pronoun clarity.
   - **Tip:** Review basic grammar rules before the test, and practice identifying errors in sentences. Familiarity with these rules will make it easier to spot mistakes.
3. **Pay Attention to Transition Words:**
   - Questions that ask you to choose the best transition between sentences or paragraphs require you to understand the logical flow of ideas.
   - **Tip:** Look for transition words like "however," "therefore," and "additionally" to determine how the sentences or ideas are connected.
4. **Maintain Consistent Tone and Style:**
   - Ensure that the tone and style of the passage remain consistent. If a passage is formal, avoid answer choices that introduce casual or informal language.
   - **Tip:** Consider the audience and purpose of the passage when choosing your answer. The correct choice will always match the overall tone of the passage.
5. **Use the Process of Elimination:**
   - When you're unsure about a question, eliminate choices that are clearly incorrect or don't fit the context. This increases your chances of selecting the correct answer.
   - **Tip:** If two answers are similar, there's a good chance one of them is correct. Compare them carefully to determine which one better fits the context.
6. **Read the Entire Sentence:**
   - For questions that focus on a specific part of a sentence, make sure to read the entire sentence or even the surrounding sentences to fully understand the context.
   - **Tip:** Don't just focus on the underlined portion; how it fits into the larger sentence or paragraph is key to determining the correct answer.
7. **Practice Revising Sentences for Style and Logic:**
   - Some questions require you to revise sentences to improve their clarity, logic, or style. Focus on making sentences clearer and more effective.
   - **Tip:** When revising, prioritize the answer that improves sentence flow and makes the point more directly without adding unnecessary information.

Final Thoughts

Consistent practice and familiarity with the types of questions you will encounter are crucial to success in the SAT Reading and Writing sections. Use these tips and tricks to refine your approach, manage your time effectively, and improve your accuracy. With the right strategies, you can boost your confidence and your score on test day.

Section 3: Essay

Essay Test Overview

The Essay section of the SAT is optional, but it offers students the opportunity to demonstrate their ability to analyze a text and craft a coherent, well-structured response. While not all colleges require the SAT Essay, submitting an essay score can provide additional insight into your writing and analytical skills, making your application stand out.

Structure and Timing:

- **Time Allotted:** 50 minutes
- **Task:** Analyze a passage provided in the test and write an essay that explains how the author builds an argument to persuade their audience.
- **Length:** The essay should be approximately 650–750 words, but there is no strict word limit. Focus on depth of analysis and clarity of expression.

What You'll Be Asked to Do:

You will be given a passage of approximately 650–750 words that presents an argument on a specific topic. Your task is to analyze how the author constructs this argument, considering elements such as the use of evidence, reasoning, and stylistic or persuasive techniques. You are not required to agree or disagree with the author's perspective; instead, your essay should focus on how effectively the author makes their case.

Skills Assessed:

1. **Reading:** Comprehend the passage and identify the author's central argument and supporting points.
2. **Analysis:** Evaluate the methods the author uses to build their argument, including the use of evidence, logical reasoning, and rhetorical strategies.
3. **Writing:** Develop a clear and cohesive essay that presents your analysis in an organized and effective manner. Use precise language and maintain a formal, academic tone.

Scoring Criteria:

The SAT Essay is scored on three criteria:

1. **Reading (1-4 points):** Assesses your understanding of the passage.
2. **Analysis (1-4 points):** Evaluates your ability to analyze how the author builds their argument.
3. **Writing (1-4 points):** Judges the quality of your writing, including organization, coherence, and language use.

Each criterion is scored by two different readers, with scores ranging from 2 to 8 points per category.

Why the Essay Matters:

While the Essay is optional, it can be a valuable component of your SAT score if you are applying to schools that require or recommend it. Even if it's not required, a strong essay score can demonstrate your writing and critical thinking abilities, particularly if these skills are a significant focus of the programs you are applying to.

Key Takeaways:

- **Understand the Task:** Focus on how the author constructs their argument, rather than your personal opinion on the topic.
- **Practice Timed Writing:** Develop your ability to write under timed conditions to ensure you can complete your essay within the 50-minute limit.
- **Focus on Analysis:** Prioritize your analysis of the author's methods, rather than summarizing the passage.
- **Use Evidence from the Passage:** Support your analysis with direct references to the text.

Scoring Criteria

Understanding how the SAT Essay is scored can help you focus on the key elements that will make your essay stand out. The SAT Essay is scored on three criteria: Reading, Analysis, and Writing. Each criterion is scored by two different readers on a scale of 1 to 4, with the total score ranging from 2 to 8 for each criterion. Here's a detailed breakdown of what the scorers are looking for in each category:

1. Reading (1-4 points):

What It Assesses:

- Your ability to comprehend the passage.
- Your understanding of the author's central argument, purpose, and key details.

Key Elements:

- **Comprehension:** Demonstrate a clear understanding of the passage by accurately identifying the author's main points and supporting evidence.
- **Central Argument:** Show that you grasp the author's primary claim or argument.
- **Evidence and Details:** Reference specific examples from the text to illustrate your understanding.

Tips for a High Score:

- **Thorough Reading:** Take the time to carefully read and understand the passage before you begin writing.
- **Accurate Summary:** When summarizing the author's argument in your essay, ensure that you accurately convey the main idea and key points.
- **Use Direct Quotes Sparingly:** While direct quotes can be helpful, ensure they are relevant and support your analysis rather than simply padding your essay.

2. Analysis (1-4 points):

What It Assesses:

- Your ability to analyze how the author builds their argument.
- Your evaluation of the use of evidence, reasoning, and rhetorical techniques.

Key Elements:

- **Argument Structure:** Identify and analyze the techniques the author uses to construct their argument, such as logical reasoning, emotional appeals, or ethical appeals (ethos, pathos, logos).
- **Evidence:** Evaluate the effectiveness of the evidence the author provides to support their claims.
- **Rhetorical Strategies:** Discuss how the author uses rhetorical devices, such as analogies, repetition, or rhetorical questions, to persuade the audience.

Tips for a High Score:

- **Deep Analysis:** Go beyond surface-level observations. Analyze how specific techniques contribute to the effectiveness of the author's argument.
- **Focus on Key Techniques:** Identify the most prominent rhetorical strategies used in the passage, and explain why they are effective in persuading the audience.
- **Connect Techniques to Purpose:** Link the author's techniques directly to their purpose in writing the passage, showing how each method supports the overall argument.

3. Writing (1-4 points):

What It Assesses:

- Your ability to craft a well-organized, coherent, and clear essay.
- Your use of language, including grammar, syntax, and word choice.

Key Elements:

- **Organization:** Structure your essay with a clear introduction, body paragraphs, and conclusion. Ensure that each paragraph flows logically from one to the next.
- **Clarity and Precision:** Use clear and concise language. Avoid vague or ambiguous statements.

- **Formal Tone:** Maintain a formal, academic tone throughout your essay. Avoid colloquialisms or overly casual language.
- **Grammar and Syntax:** Use correct grammar, punctuation, and sentence structure. Vary your sentence lengths and structures to create a more engaging and dynamic essay.

Tips for a High Score:

- **Outline Before Writing:** Spend a few minutes planning your essay structure before you begin writing. This will help you stay organized and focused.
- **Strong Thesis Statement:** Start with a clear thesis statement that outlines your analysis of the author's argument.
- **Effective Transitions:** Use transition words and phrases to ensure a smooth flow between paragraphs and ideas.
- **Proofread:** If time allows, quickly review your essay for any grammatical or syntactical errors before submitting.

Final Scoring Insights:

- **Balanced Scores:** Aim for a balanced performance across all three criteria. A strong essay will demonstrate a solid understanding of the passage, a thoughtful analysis of the author's techniques, and clear, effective writing.
- **Practice:** Regular practice with timed essays can help you improve your ability to perform well in all three areas. Review your practice essays critically, or seek feedback from teachers or peers to identify areas for improvement.

Essay Writing Skills

Writing a strong SAT essay requires a combination of planning, analysis, and effective writing. In this section, we'll explore the key skills you need to develop a compelling essay that meets the SAT's scoring criteria.

Planning and Organizing Your Essay

1. Analyze the Prompt:

- Begin by carefully reading the essay prompt and passage. Understand the author's main argument and the techniques they use to build their case.
- **Tip:** Spend the first few minutes of your essay time analyzing the passage and identifying the key points you'll address in your essay.

2. Develop a Thesis Statement:

- Your thesis should clearly state your analysis of how the author builds their argument. It should reflect your understanding of the passage and outline the main points you'll discuss.
- **Example Thesis:** "In the passage, the author effectively uses logical reasoning, strong evidence, and emotional appeals to persuade the audience of the urgency of environmental conservation."

3. Outline Your Essay:

- Before writing, create a brief outline to organize your thoughts. Structure your essay with an introduction, body paragraphs, and a conclusion.
- **Introduction:** Introduce the passage, briefly summarize the author's argument, and state your thesis.
- **Body Paragraphs:** Each paragraph should focus on a different technique the author uses. Start with a topic sentence, provide examples from the text, and analyze how these examples support the author's argument.
- **Conclusion:** Summarize your analysis and restate your thesis, emphasizing the overall effectiveness of the author's argument.

Developing a Strong Thesis

1. Make It Specific:

- A strong thesis statement is specific and clear. It should directly address the author's techniques and the purpose of their argument.
- **Weak Thesis:** "The author makes a good argument."
- **Strong Thesis:** "The author effectively uses statistical evidence, emotional appeals, and expert testimony to argue that stricter environmental regulations are necessary."

2. Ensure It's Debatable:

- Your thesis should present an argument that you'll support with evidence and analysis. It shouldn't be a simple statement of fact.
- **Example:** "Through the use of vivid imagery and logical reasoning, the author convinces the audience of the pressing need for climate action."

Using Evidence Effectively

1. Cite Specific Examples:

- Support your analysis with direct references to the passage. Use specific examples of the author's use of evidence, reasoning, and rhetorical devices.
- **Example:** "The author cites a study showing a 40% increase in pollution levels over the past decade to emphasize the urgency of environmental reform."

2. Explain the Impact:

- Don't just list examples—explain how they contribute to the author's overall argument. Analyze the effectiveness of the evidence and why it persuades the audience.
- **Example:** "By including this statistic, the author appeals to the audience's logical side, providing concrete evidence of the worsening environmental crisis."

3. Avoid Over-Quoting:

- Use quotations sparingly. Instead of quoting large sections of the text, paraphrase when appropriate and focus on analyzing the content.
- **Tip:** Quotations should only be used when the exact wording is crucial to your analysis.

Crafting Strong Body Paragraphs

1. Start with a Topic Sentence:

- Each body paragraph should begin with a clear topic sentence that introduces the main idea of the paragraph and links back to your thesis.
- **Example:** "One of the key strategies the author employs is the use of logical reasoning to build a convincing argument."

2. Provide Analysis, Not Summary:

- Your essay should focus on analyzing how the author builds their argument, rather than summarizing the passage. Discuss the "how" and "why" of the author's techniques.
- **Example:** "The author's use of logical reasoning, particularly in the presentation of statistics and expert opinions, strengthens the credibility of the argument and makes it more compelling."

3. Use Transition Words:

- Transition words help your essay flow smoothly from one point to the next. Use them to connect ideas within and between paragraphs.
- **Examples:** "Furthermore," "Moreover," "In addition," "However," "Therefore."

Writing a Coherent Conclusion

1. Restate Your Thesis:

- Begin your conclusion by restating your thesis in a new way. Summarize the main points you made in your essay.

- **Example:** "In conclusion, through the effective use of logical reasoning, compelling evidence, and emotional appeals, the author successfully persuades the audience of the necessity of immediate environmental action."

2. Reflect on the Overall Impact:

- Discuss the overall impact of the author's argument. Why is it effective? What makes the argument persuasive or compelling?
- **Example:** "The author's strategic use of evidence and rhetorical devices not only highlights the severity of the issue but also inspires the audience to consider their role in addressing it."

3. End with a Strong Closing Statement:

- Finish your essay with a strong, memorable closing statement that leaves a lasting impression on the reader.
- **Example:** "Ultimately, the author's argument serves as a powerful call to action, urging readers to take responsibility for the future of our planet."

Final Writing Tips:

- **Practice Under Timed Conditions:** Regular practice will help you get comfortable with the time constraints of the SAT Essay. Set aside 50 minutes and write practice essays to build your confidence.
- **Review and Revise:** If time permits, quickly review your essay for any errors or areas that could be improved. A few minutes of revision can make a significant difference in your final score.
- **Stay Focused:** Keep your essay focused on analyzing the passage. Avoid bringing in unrelated information or opinions.

Sample Prompts and High-Scoring Essays

In this section, we will explore sample prompts that mirror the types of passages you might encounter on the SAT Essay. Additionally, we'll provide examples of high-scoring essays with explanations of what makes them effective. Analyzing these examples will help you understand what the SAT essay scorers are looking for and how to structure your own essay for maximum impact.

Sample Prompt 1

Prompt: As you read the passage below, consider how the author uses evidence, such as facts or examples, to support claims, reasoning to develop ideas and to connect claims and evidence, and stylistic or persuasive elements, such as word choice or appeals to emotion, to add power to the ideas expressed.

Passage Excerpt: "Today, more than ever, we are witnessing the devastating effects of climate change across the globe. Rising sea levels, more frequent and severe natural disasters, and shifts in weather patterns are just a few examples of the impact of human activity on the environment. To combat this growing crisis, we must take immediate and decisive action. This includes reducing greenhouse gas emissions, transitioning to renewable energy sources, and implementing policies that prioritize environmental sustainability. The future of our planet depends on the actions we take today."

High-Scoring Essay Example:

Essay:

In the passage, the author effectively builds an argument to persuade the audience of the urgent need for action to address climate change. By employing a combination of logical reasoning, compelling evidence, and emotional appeals, the author creates a strong and persuasive case that emphasizes the severity of the issue and the importance of immediate action.

The author begins by presenting a series of alarming statistics and examples, such as rising sea levels and more frequent natural disasters, to illustrate the devastating effects of climate change. This use of evidence is particularly effective because it provides concrete examples that are difficult to dismiss, making the argument more credible. The author's reliance on factual data not only strengthens the argument but also appeals to the audience's sense of logic and reason.

In addition to logical reasoning, the author uses emotional appeals to connect with the audience on a deeper level. By discussing the impact of climate change on the future of the planet, the author taps into the audience's fear and concern for the well-being of future generations. This appeal to emotion is powerful because it makes the issue more personal and urgent, encouraging the audience to take action.

Finally, the author's use of stylistic elements, such as the repetition of the phrase "we must," serves to reinforce the call to action. The repetition creates a sense of urgency and collective responsibility, reminding the audience that the fight against climate change is a shared effort. This rhetorical strategy is effective in motivating the audience to consider their own role in addressing the issue.

Overall, the author's argument is both compelling and persuasive. Through the strategic use of evidence, reasoning, and rhetorical techniques, the author successfully convinces the audience of the importance of taking immediate and decisive action to combat climate change.

Score Breakdown:

- **Reading:** 4/4 - The essay demonstrates a thorough understanding of the passage.
- **Analysis:** 4/4 - The essay offers a detailed and insightful analysis of how the author builds the argument.
- **Writing:** 4/4 - The essay is well-organized, clearly written, and free of significant errors.

Sample Prompt 2

Prompt: Consider how the author uses evidence, such as facts or examples, reasoning to develop ideas and to connect claims and evidence, and stylistic or persuasive elements to strengthen their argument.

Passage Excerpt: "In an age where technology is rapidly advancing, it is crucial to address the ethical implications of artificial intelligence (AI). While AI offers numerous benefits, such as increased efficiency and the ability to process vast amounts of data, it also poses significant ethical challenges. These include concerns about privacy, job displacement, and the potential for AI to be used in harmful ways. To ensure that AI is developed and deployed responsibly, we must establish ethical guidelines and regulations that prioritize human welfare."

High-Scoring Essay Example:

Essay:

In this passage, the author effectively argues for the need to establish ethical guidelines for the development and use of artificial intelligence (AI). Through a combination of logical reasoning, ethical considerations, and persuasive language, the author presents a compelling case for why it is essential to address the ethical implications of AI.

The author begins by acknowledging the benefits of AI, such as increased efficiency and the ability to process vast amounts of data. This balanced approach is effective because it demonstrates that the author is not opposed to technological advancement, but rather concerned with ensuring that AI is used responsibly. By presenting both the positive and negative aspects of AI, the author establishes credibility and appeals to the audience's sense of logic.

Next, the author addresses the ethical challenges posed by AI, including concerns about privacy, job displacement, and the potential for misuse. The use of specific examples, such as the loss of jobs due to automation, helps to illustrate the real-world impact of these ethical concerns. This approach is persuasive because it connects abstract ethical issues to tangible outcomes, making the argument more relatable and compelling.

The author further strengthens the argument by using persuasive language that emphasizes the importance of prioritizing human welfare in the development of AI. Phrases such as "crucial to address" and "responsibly deployed" highlight the urgency of the issue and appeal to the audience's sense of moral responsibility. This use of language is effective in motivating the audience to consider the ethical implications of AI and support the establishment of guidelines and regulations.

Overall, the author successfully builds a persuasive argument that emphasizes the importance of addressing the ethical implications of AI. Through the use of logical reasoning, specific examples, and persuasive language, the author convinces the audience of the need for responsible AI development and regulation.

Score Breakdown:

- **Reading:** 4/4 - The essay demonstrates a strong understanding of the passage.
- **Analysis:** 4/4 - The essay provides a thorough and nuanced analysis of the author's argument.
- **Writing:** 4/4 - The essay is well-structured, articulate, and free of major grammatical errors.

Key Takeaways from High-Scoring Essays:

1. **Strong Thesis Statements:** Both essays feature clear, specific thesis statements that outline the main points of analysis. A strong thesis is crucial for guiding the reader through your essay.
2. **Detailed Analysis:** High-scoring essays go beyond summarizing the passage. They provide detailed analysis of how the author uses evidence, reasoning, and rhetorical strategies to build their argument.
3. **Effective Use of Evidence:** The essays use specific examples from the passage to support their analysis. Direct references to the text strengthen the argument and demonstrate a deep understanding of the material.
4. **Coherent Structure:** Both essays are well-organized, with clear introductions, body paragraphs, and conclusions. Each paragraph has a clear focus and contributes to the overall argument.
5. **Formal and Precise Language:** The language in these essays is formal, precise, and appropriate for the task. Word choice and sentence structure are varied and contribute to the clarity of the analysis.

Practice Essays with Feedback

In this section, you'll find practice essay prompts along with example student essays and detailed feedback. This will help you understand how to improve your own writing and meet the SAT Essay scoring criteria. Each practice essay will include feedback on Reading, Analysis, and Writing to help you identify areas for improvement.

Practice Prompt 1

Prompt: Read the following passage and write an essay that analyzes how the author builds an argument to persuade the audience of the need for stricter environmental regulations.

Passage Excerpt: "Our planet is facing an unprecedented environmental crisis. The rapid industrialization of the past century has led to widespread pollution, deforestation, and the depletion of natural resources. Without immediate action, these trends will continue to accelerate, leading to irreversible damage to our ecosystems. Stricter environmental regulations are essential to mitigating these effects and ensuring a sustainable future. Governments must take bold steps to limit emissions, protect endangered species, and promote renewable energy sources. The time to act is now, before it is too late."

Student Essay Example:

Essay:

The author of the passage argues that stricter environmental regulations are necessary to combat the ongoing environmental crisis. To build this argument, the author employs a combination of logical reasoning, emotional appeals, and strong evidence. These rhetorical strategies effectively persuade the reader of the urgency and importance of taking immediate action.

One of the primary techniques the author uses is logical reasoning. The author begins by outlining the consequences of rapid industrialization, such as pollution, deforestation, and resource depletion. By presenting these negative outcomes as the result of human activity, the author makes a logical case for why stricter regulations are needed. The cause-and-effect relationship between industrialization and environmental degradation is clear and compelling.

In addition to logical reasoning, the author appeals to the reader's emotions by emphasizing the potential consequences of inaction. Phrases like "irreversible damage" and "before it is too late" are designed to evoke a sense of urgency and fear. This emotional appeal is effective because it makes the reader feel personally invested in the issue. The idea that future generations could suffer as a result of our failure to act is particularly powerful and motivates the reader to support stricter regulations.

Finally, the author supports their argument with strong evidence. The mention of specific actions, such as limiting emissions and promoting renewable energy, adds credibility to the argument. These are concrete solutions that have been proven effective in reducing environmental harm. By providing evidence-based solutions, the author strengthens their case and makes it more difficult for the reader to dismiss the need for stricter regulations.

Overall, the author successfully builds a persuasive argument for stricter environmental regulations. Through the use of logical reasoning, emotional appeals, and evidence, the author convinces the reader of the need to take immediate action to protect our planet.

Feedback:

- **Reading (Score: 3/4):** The essay demonstrates a solid understanding of the passage, but it could go deeper in analyzing the author's central argument and the evidence provided. While the main points are identified, the analysis could benefit from a more nuanced exploration of the author's use of evidence.
- **Analysis (Score: 3/4):** The analysis is clear and well-structured, but it tends to summarize rather than deeply analyze the rhetorical strategies. To achieve a higher score, the essay should delve more into how each strategy contributes to the overall persuasiveness of the argument.
- **Writing (Score: 4/4):** The essay is well-organized and clearly written, with strong transitions between paragraphs. The language is formal and precise, with varied sentence structure and no significant grammatical errors.

Practice Prompt 2

Prompt: Analyze how the author uses evidence and reasoning to argue that technology should be used to enhance, rather than replace, human labor.

Passage Excerpt: "As technology continues to advance, there is growing concern that machines will eventually replace human workers. However, rather than viewing technology as a threat, we should see it as a tool to enhance human labor. By automating repetitive tasks, technology can free up workers to focus on more creative and complex aspects of their jobs. This not only improves productivity but also increases job satisfaction. The key is to use technology in a way that complements human skills rather than replaces them."

Student Essay Example:

Essay:

In this passage, the author argues that technology should be used to enhance human labor rather than replace it. The author builds this argument by using logical reasoning and evidence to demonstrate the benefits of integrating technology into the workplace. These strategies are effective in convincing the reader that technology can be a positive force when used correctly.

The author begins by addressing the common concern that machines will replace human workers. By acknowledging this fear, the author establishes common ground with the reader. The author then counters this concern by presenting the idea that technology can enhance human labor by automating repetitive tasks. This logical reasoning is effective because it reframes the issue, showing that technology can actually improve the quality of work rather than diminish it.

The author also uses evidence to support their argument. For example, the author mentions that automation can increase productivity and job satisfaction by allowing workers to focus on more complex tasks. This evidence is persuasive because it is based on real-world examples of how technology is being used in the workplace. By providing concrete benefits, the author strengthens their case and makes it more convincing.

Additionally, the author emphasizes that the key to successfully integrating technology is to use it in a way that complements human skills. This idea is reinforced by the author's use of language, such as the phrase "complements human skills." This choice of words suggests that technology and human labor can work together in harmony, rather than being in competition with one another.

Overall, the author effectively argues that technology should be used to enhance human labor. Through the use of logical reasoning, evidence, and persuasive language, the author convinces the reader that technology can be a valuable tool when used appropriately.

Feedback:

- **Reading (Score: 4/4):** The essay demonstrates a thorough understanding of the passage. The student accurately identifies the author's argument and key points, showing a deep comprehension of the material.
- **Analysis (Score: 3/4):** The analysis is solid but could be more detailed. The essay identifies the rhetorical strategies used by the author, but it could delve further into how these strategies effectively persuade the reader. For instance, discussing the impact of the author's language choices on the audience could strengthen the analysis.
- **Writing (Score: 4/4):** The essay is well-organized and clearly written, with a strong introduction, body paragraphs, and conclusion. The language is formal and appropriate for the task, with no significant grammatical errors.

Key Feedback Themes:

1. **Depth of Analysis:** High-scoring essays go beyond summarizing the passage. Focus on deeply analyzing how the author's use of evidence, reasoning, and rhetorical techniques contributes to the overall persuasiveness of the argument.
2. **Use of Evidence:** When discussing the author's use of evidence, be specific about how it supports the argument. Explain why the evidence is persuasive and how it impacts the reader.

3. **Language and Tone:** Maintain a formal, academic tone throughout your essay. Avoid casual language, and ensure that your writing is clear and precise.
4. **Structure and Organization:** Strong essays have a clear structure, with a well-defined introduction, body paragraphs, and conclusion. Use topic sentences and transitions to guide the reader through your analysis.

Section 4: Test-Taking Strategies

Effective test-taking strategies can make a significant difference in your SAT performance. By approaching the test with the right mindset and techniques, you can maximize your score and reduce test-day stress. This section will guide you through strategies for preparing effectively, managing your time, and tackling different types of questions.

General Test-Taking Strategies

1. Effective Study Habits

Developing effective study habits is crucial for success on the SAT. Consistency and a well-structured study plan will help you cover all the material and build confidence.

- **Create a Study Schedule:** Establish a regular study routine leading up to the test. Break your study sessions into manageable chunks, focusing on different sections of the SAT each day. For example, you might dedicate Mondays to Math, Wednesdays to Reading, and Fridays to Writing and Language.
- **Use Quality Study Materials:** Invest in reputable SAT prep books, online resources, and practice tests. The College Board offers official practice tests that closely resemble the actual SAT, making them a valuable tool in your preparation.
- **Practice with Timed Tests:** Simulate test-day conditions by taking full-length, timed practice tests. This will help you get used to the pacing of the exam and identify areas where you need to improve your speed or accuracy.
- **Review Your Mistakes:** After each practice test or study session, review the questions you got wrong. Understand why you made each mistake and focus on improving in those areas. This targeted practice will help you avoid making the same errors on test day.
- **Stay Organized:** Keep track of your progress by maintaining a study journal or checklist. Note the areas where you struggle and adjust your study plan accordingly.
- **Take Care of Your Health:** Ensure you're getting enough sleep, eating well, and staying hydrated, especially in the days leading up to the test. Your physical well-being directly impacts your mental performance.

2. Time Management During the Test

Managing your time effectively during the SAT is essential to completing all sections of the test without feeling rushed. Here are some strategies to help you stay on track:

- **Pace Yourself:** Each section of the SAT is timed, so it's important to keep an eye on the clock. For example, you have 65 minutes to complete the Reading section, which consists of 52 questions. This means you should spend roughly 75 seconds per question. If you find yourself spending too much time on one question, move on and return to it later if you have time.
- **Use Your Breaks Wisely:** The SAT includes a few short breaks. Use these breaks to relax, clear your mind, and prepare for the next section. Stretch your legs, take deep breaths, and hydrate. Avoid thinking about the previous section—focus on what's coming next.
- **Prioritize Easy Questions:** Answer the questions you find easiest first, especially in sections like Math. This will help you secure points early on and build confidence. Return to the more challenging questions with the time you have left.
- **Watch for Time Traps:** Some questions, particularly in the Reading and Math sections, can be more time-consuming than others. Be aware of these potential time traps and avoid getting stuck on them. If a question is taking too long, make an educated guess, mark it, and move on.
- **Use the Process of Elimination:** If you're unsure of the answer, eliminate the choices you know are incorrect. This increases your chances of selecting the correct answer from the remaining options.
- **Keep Moving:** Don't dwell too long on any single question. It's better to guess and move on than to run out of time with unanswered questions. Remember, there's no penalty for guessing on the SAT.

Guessing Strategies

When taking the SAT, it's important to have a strategy for dealing with questions you find difficult or are unsure about. Guessing strategically can help you maximize your score, especially since there's no penalty for wrong answers on the SAT. Here are some effective guessing strategies:

Guessing Strategies

1. Eliminate Wrong Answers

One of the most effective guessing strategies is the process of elimination. By narrowing down the answer choices, you increase your chances of selecting the correct answer, even if you're unsure.

- **Look for Obvious Mistakes:** Start by eliminating any answer choices that are clearly incorrect. For example, if a math question involves positive numbers, you can eliminate any negative answer choices.
- **Watch for Extreme Words:** In the Reading and Writing sections, answer choices with extreme words like "always," "never," or "only" are often incorrect. SAT questions tend to favor more moderate or nuanced answers.
- **Consider the Context:** For vocabulary in context questions, eliminate any choices that don't fit the overall meaning of the sentence or passage. Sometimes, the answer choice that sounds "fancier" is not the correct one.

2. Make an Educated Guess

If you've eliminated one or more wrong answers but still aren't sure of the correct one, make an educated guess from the remaining options.

- **Look for Patterns:** Sometimes the SAT will present patterns or consistencies in correct answers. For example, if two answer choices are similar, they may be more likely to be correct than choices that are completely different.
- **Use the Question Stem:** Pay close attention to the wording of the question stem, as it often contains clues about the correct answer. For instance, if the question asks for the "best" or "most effective" solution, focus on answers that directly address the main issue.
- **Trust Your Gut:** If you're torn between two choices, go with your first instinct. Often, your initial reaction is based on an intuitive understanding of the material.

3. Random Guessing When Time Is Running Out

If you're running out of time and haven't finished a section, it's better to fill in answers randomly than to leave questions blank.

- **Use a Consistent Letter:** If you have to guess on several questions in a row, choose a consistent letter (e.g., always pick "B") rather than randomly selecting different letters. Statistically, this approach may improve your chances of getting at least some questions correct.
- **Leave Time for Random Guessing:** Aim to complete the test with a few minutes to spare, so you can quickly guess on any unanswered questions. Use the remaining time to review and make any final changes.

4. Mark and Return to Difficult Questions

During the test, if you encounter a particularly difficult question, mark it and move on to the next one. This prevents you from wasting time and allows you to come back to it with a fresh perspective.

- **Don't Dwell on Uncertain Questions:** Spending too much time on one question can cost you valuable points elsewhere. Move on and return to it if time allows.
- **Reassess with Fresh Eyes:** Sometimes returning to a difficult question after answering others can help you see it differently. You may catch something you missed the first time.

5. Avoid Second-Guessing Yourself

While it's important to review your answers if time permits, be cautious about changing your answers too often.

- **Trust Your First Answer:** Studies show that first instincts are often correct. Change your answer only if you're sure you made a mistake or if new information has come to light.
- **Review with Focus:** When reviewing, focus on questions you were unsure about or skipped. Don't second-guess yourself on questions you felt confident about.

Answer Elimination Techniques

Answer elimination techniques are essential tools for navigating tricky SAT questions, particularly when you're unsure of the correct answer. By systematically eliminating incorrect options, you increase your odds of selecting the right answer. Here's how you can effectively use these techniques across different sections of the SAT:

Answer Elimination Techniques

1. Identify and Eliminate Extreme or Absolute Answers

In the Reading and Writing sections, answers that contain extreme language or absolutes are often incorrect. Words like "always," "never," "all," or "none" can signal an answer choice that is too rigid to be correct, especially in questions that require nuance.

- **Look for Moderate Options:** SAT questions often favor answers that allow for exceptions or reflect more balanced viewpoints. For instance, an answer choice that includes words like "often," "sometimes," or "generally" may be more likely to be correct.
- **Apply This to Vocabulary in Context:** When asked to determine the meaning of a word in context, eliminate any choices that seem too extreme given the sentence or passage.

2. Use Logic to Eliminate Implausible Choices

Some answer choices may simply not make sense within the context of the question. Use your reasoning skills to identify these implausible options and eliminate them quickly.

- **Match the Answer to the Question:** Ensure that the answer choice directly addresses the question being asked. If an answer choice goes off-topic or doesn't make logical sense in relation to the question, it's likely incorrect.
- **In Math Sections:** Eliminate answers that don't fit the parameters of the problem. For example, if the problem involves calculating the length of a side in a triangle, eliminate answers that are negative or don't conform to the rules of geometry.

3. Cross-Check with the Passage

In the Reading section, you can often eliminate incorrect answers by cross-checking them against the passage. If an answer choice contradicts or is not supported by the passage, you can safely eliminate it.

- **Find Textual Evidence:** Refer back to the passage to verify the accuracy of an answer choice. If the passage doesn't support the claim made by an answer, it's not the right choice.
- **Eliminate Paraphrasing Traps:** Some incorrect choices paraphrase part of the passage in a misleading way. Be wary of answers that seem close to the text but distort its meaning.

4. Look for Contradictions in the Math Section

In the Math sections, incorrect answers often contradict the conditions set by the problem. Carefully check that the answer choice satisfies all given conditions.

- **Verify Units and Dimensions:** Ensure that the units in the answer choice match what's being asked for in the problem. If a problem asks for an answer in square units, eliminate choices that are in linear units or don't match the expected format.
- **Plug in Values:** For algebraic problems, you can sometimes eliminate choices by plugging in values from the answer choices into the original equation or inequality to see which ones hold true.

5. Consider the Question Stem

The wording of the question stem often provides clues about what makes an answer choice correct or incorrect. Pay close attention to key phrases in the question.

- **Match Scope and Focus:** Ensure that the answer choice is neither too broad nor too narrow in relation to the question stem. If the question is about a specific detail, eliminate answers that are too general.
- **Avoid Off-Topic Choices:** Sometimes, an answer may seem plausible on its own but doesn't directly answer the question posed in the stem. These off-topic choices can be safely eliminated.

6. Use Patterns in Answer Choices

Occasionally, answer choices follow certain patterns that can help you eliminate incorrect options. For example, two choices may be similar except for a minor detail, indicating that one of them is likely correct.

- **Identify Redundant Choices:** If two answer choices are nearly identical, the correct answer is often one of these. Look closely at the differences to determine which one aligns better with the question.
- **Beware of Decoy Choices:** Sometimes, a question will include a choice that is a common misconception or a trap answer designed to mislead you. If a choice seems too obvious or simplistic, it might be worth a second look.

7. Revisit Eliminated Choices

If you're left with a difficult decision between two or three choices, revisit the ones you initially eliminated. With the process of elimination, you might reconsider them and find that one of them fits better than your remaining choices.

- **Double-Check Your Work:** When reviewing eliminated choices, ensure that your reasoning for eliminating them was sound. Sometimes, a second look reveals an overlooked detail that could make a previously eliminated choice correct.
- **Consider the Lesser Evil:** If you're genuinely stuck, and none of the remaining choices seem ideal, choose the one that is "least wrong" or aligns most closely with the evidence.

By mastering these answer elimination techniques, you can tackle even the toughest SAT questions with greater confidence. These strategies will help you narrow down your options and make more informed guesses, ultimately improving your overall score.

Techniques for Narrowing Down Multiple-Choice Options

When faced with multiple-choice questions, particularly those that are challenging or confusing, it's essential to use effective techniques to narrow down the options and increase your chances of selecting the correct answer. Here's how you can systematically approach multiple-choice questions to improve your accuracy:

Techniques for Narrowing Down Multiple-Choice Options

1. Rephrase the Question in Your Own Words

Before looking at the answer choices, try to rephrase the question in your own words. This helps you focus on what the question is actually asking and can prevent you from being misled by tricky wording.

- **Clarify the Question:** If the question is complex, break it down into simpler parts. Ask yourself what the key point is and what the question is really seeking.
- **Example:** If the question asks, "Which choice best supports the author's argument in the passage?" rephrase it to, "Which answer shows evidence that strengthens what the author is saying?"

2. Predict the Answer Before Viewing Choices

After rephrasing the question, try to predict the answer before looking at the provided options. This allows you to approach the choices with a clear idea of what you're looking for, making it easier to identify the correct answer.

- **Use Context Clues:** For Reading and Writing questions, use the surrounding sentences or paragraphs to anticipate the answer.
- **Example:** If the passage suggests a particular outcome or opinion, predict what that might be before comparing it to the choices.

3. Compare All Choices Before Selecting an Answer

Even if one of the first answer choices seems correct, it's important to read through all the options. Sometimes a later choice might be better or more complete.

- **Consider Each Choice:** Evaluate each answer in relation to the question. Compare them to your predicted answer and see which one aligns best.
- **Avoid Jumping to Conclusions:** Don't rush to select an answer without fully considering all the alternatives.

4. Look for the Most Specific and Relevant Answer

In many cases, the most specific and directly relevant answer is the correct one. Vague or overly broad answers are often incorrect, especially when the question is asking for detailed information.

- **Match Specificity:** Choose the answer that directly addresses the question with precise details, rather than one that is general or off-topic.
- **Example:** If a question asks for the main idea of a paragraph, select the answer that encapsulates the key points rather than one that makes a broad or unrelated statement.

5. Eliminate Choices That Are Only Partially Correct

Sometimes, answer choices will contain elements that are correct but overall are not the best choice. Look for options that are partially right but don't fully answer the question.

- **Identify Partial Answers:** If a choice seems somewhat correct but doesn't fully address the question or has extraneous information, eliminate it.
- **Example:** In a Math problem, an answer that correctly solves part of the equation but ignores other important aspects should be eliminated.

6. Be Wary of Answers That Are Too Similar

When two or more answer choices are very similar, there's a good chance that one of them is correct. However, be cautious and carefully compare the differences.

- **Spot the Key Difference:** Determine what makes the similar choices different from each other and which difference aligns best with the question.
- **Example:** If two answers are the same except for a single word or phrase, focus on that detail to decide which is correct.

7. Consider the Tone and Style in Reading and Writing Sections

For questions in the Reading and Writing sections, pay attention to the tone and style of the passage. The correct answer should match the tone and intent of the original text.

- **Match the Author's Intent:** Choose the answer that fits the author's tone—whether it's formal, informal, persuasive, or descriptive.
- **Example:** If the passage is written in a formal tone, eliminate any answer choices that use casual language.

8. Apply the Rule of Consistency in Math

In the Math section, ensure that your answer is consistent with the information provided in the problem. Inconsistent units, incorrect operations, or illogical outcomes are signs of wrong answers.

- **Check for Logical Consistency:** Ensure that the solution aligns with the given data and the logical requirements of the problem.
- **Example:** If the problem deals with dimensions, eliminate answers that provide inconsistent units or results that don't make sense in the context.

9. Avoid Choosing Answers Based on Familiarity

Sometimes, test-takers choose answers that seem familiar or similar to something they've studied, even if it doesn't fit the context of the question. Avoid this trap by focusing on what the question is asking, not on what you remember from your studies.

- **Focus on Context:** Make sure the answer fits within the specific context of the question rather than just sounding familiar.
- **Example:** If a question is about a specific event in history, don't choose an answer just because it's a famous event—choose the one that directly answers the question.

10. Use the Process of Elimination Efficiently

The process of elimination is most effective when you systematically remove incorrect answers, leaving you with fewer options to consider.

- **Eliminate Quickly:** As soon as you identify an incorrect option, cross it out or mentally eliminate it. This reduces the number of choices and increases your focus on the remaining options.
- **Trust the Process:** If you've narrowed it down to two choices and are unsure, pick the one that seems more consistent with the question and your analysis.

By mastering these techniques for narrowing down multiple-choice options, you can approach each question with greater confidence and precision. These strategies will help you eliminate incorrect answers and choose the best possible option, ultimately leading to a higher score on the SAT.

Practice Scenarios

To solidify your understanding of test-taking strategies and to help you apply them effectively, let's walk through a few practice scenarios. These scenarios will simulate the types of questions you might encounter on the SAT and show how to apply the strategies we've discussed.

Scenario 1: Reading Section - Identifying the Main Idea

Question: In the passage, the author argues that the rise of digital media has led to a decline in traditional print journalism. Which of the following statements best captures the main idea of the passage?
A) Digital media offers more diverse and timely content than traditional print journalism. B) The decline of print journalism is inevitable due to the convenience and accessibility of digital media. C) Traditional print journalism still has a place in the digital age, despite the rise of online news sources. D) Digital media has replaced print journalism as the primary source of news for most people.

Applying the Strategy:

1. **Rephrase the Question:** What is the author's main point about digital media and print journalism?
2. **Predict the Answer:** The author likely argues that digital media is causing the decline of print journalism.
3. **Eliminate Wrong Answers:**
    - **A)** This answer focuses on the benefits of digital media, but doesn't address the decline of print journalism.
    - **C)** This answer suggests that print journalism still thrives, which contradicts the author's argument.
    - **D)** This answer is close, but it emphasizes replacement rather than decline.
4. **Choose the Most Relevant Answer:**
    - **B)** This choice directly addresses the decline of print journalism and matches the predicted answer.

Correct Answer: B) The decline of print journalism is inevitable due to the convenience and accessibility of digital media.

Scenario 2: Writing Section - Improving Sentence Clarity

Question: Which choice best improves the clarity and conciseness of the following sentence?

"Due to the fact that the company was facing a significant decrease in revenue, the management team decided that it was necessary to implement cost-cutting measures in order to improve profitability."

A) Due to the fact that the company was facing a significant decrease in revenue, the management team decided that it was necessary to implement cost-cutting measures in order to improve profitability. B) The company's revenue was significantly decreasing, so the management team decided to implement cost-cutting measures to improve profitability. C) Because the company was facing a decrease in revenue, the management team decided that implementing cost-cutting measures was necessary to improve profitability. D) Facing a significant revenue decrease, the management team decided to implement cost-cutting measures to improve profitability.

Applying the Strategy:

1. **Rephrase the Question:** How can the sentence be made clearer and more concise?
2. **Predict the Answer:** The sentence should be shorter and more direct, avoiding unnecessary words.
3. **Eliminate Wrong Answers:**
   - **A)** This is the original sentence, which is wordy and repetitive.
   - **C)** While this option removes some redundancy, it still includes unnecessary words like "that."
4. **Choose the Most Concise and Clear Answer:**
   - **B)** This option simplifies the sentence but is slightly less direct than **D.**
   - **D)** This choice is the most concise and clear, directly addressing the issue with the fewest words.

Correct Answer: D) Facing a significant revenue decrease, the management team decided to implement cost-cutting measures to improve profitability.

Scenario 3: Math Section - Solving Word Problems

Question: A car rental company charges a daily rate of $50 plus $0.20 per mile driven. If a customer's total bill was $130, how many miles did they drive?

A) 200 miles B) 300 miles C) 400 miles D) 500 miles

Applying the Strategy:

1. **Rephrase the Question:** How many miles did the customer drive if their total bill was $130?
2. **Set Up the Equation:**
   - Let x represent the number of miles driven.
   - The total cost is given by the equation: $50 + 0.20x = 130$.
3. **Solve the Equation:**
   - Subtract $50 from both sides: $0.20x = 80$.
   - Divide both sides by $0.20: $x = 400$.
4. **Check the Answer Choices:**
   - The correct answer is **400 miles.**

Correct Answer: C) 400 miles

These practice scenarios illustrate how to apply the test-taking strategies we've discussed. By following these steps, you can approach each question systematically and improve your chances of selecting the correct answer.

Stress and Anxiety Management

Test anxiety is a common challenge for many students, but with the right strategies, you can manage stress and stay calm during the SAT. This section will provide you with techniques to help reduce anxiety, stay focused, and perform your best on test day.

Techniques to Stay Calm

1. Practice Mindfulness and Relaxation Techniques

Mindfulness and relaxation techniques can help you stay calm and focused during the test. These techniques are easy to practice and can be highly effective in managing anxiety.

- **Deep Breathing Exercises:** Practice deep breathing before and during the test. Take a slow, deep breath in through your nose, hold it for a few seconds, and then slowly exhale through your mouth. Repeat this process a few times to calm your nerves and lower your heart rate.
- **Progressive Muscle Relaxation:** Tense and then slowly relax different muscle groups in your body, starting from your toes and working your way up to your head. This technique helps release physical tension and can reduce feelings of anxiety.
- **Visualization:** Close your eyes and visualize yourself successfully completing the test. Imagine feeling calm, confident, and in control as you work through each section. Visualization can help create a positive mindset and reduce stress.

2. Focus on the Present Moment

Anxiety often arises from worrying about the future or dwelling on past mistakes. By focusing on the present moment, you can prevent your mind from wandering and stay centered on the task at hand.

- **Stay Grounded:** If you feel your mind starting to race, bring your focus back to the test by grounding yourself in the present. Look at the question in front of you and remind yourself that you only need to focus on one question at a time.
- **Use a Mantra:** Repeat a calming phrase or mantra to yourself, such as "I am prepared," "I can do this," or "One question at a time." This can help redirect your thoughts and keep you focused.

3. Break Down the Test into Manageable Parts

The SAT is a long and comprehensive test, which can be overwhelming if you think about it as a whole. Instead, break it down into smaller, more manageable parts.

- **Take It One Section at a Time:** Focus on completing each section to the best of your ability before moving on to the next. Don't worry about future sections while you're working on the current one.

- **Set Mini-Goals:** Set small, achievable goals for yourself throughout the test. For example, aim to answer a certain number of questions in a set amount of time, or focus on getting through a passage with full comprehension.

4. Prepare for Test Day in Advance

Preparation can significantly reduce anxiety because it eliminates the uncertainty that often causes stress. By preparing thoroughly, you can walk into the test knowing you've done everything possible to succeed.

- **Familiarize Yourself with the Test Format:** Know what to expect on test day by practicing with full-length SAT tests. This familiarity will make the real test feel less intimidating.
- **Plan Your Test Day Logistics:** Plan your route to the test center, know what to bring, and prepare everything you need the night before. This will help you feel more in control and reduce last-minute stress.
- **Get a Good Night's Sleep:** Ensure you get enough rest the night before the test. Sleep is crucial for cognitive function, and being well-rested will help you think clearly and stay focused during the exam.

5. Maintain Perspective

Remember that while the SAT is important, it's not the only factor that colleges consider in the admissions process. Maintaining perspective can help you manage stress and keep the test in context.

- **Know It's Just One Part of Your Application:** Colleges look at many factors beyond your SAT score, such as your GPA, extracurricular activities, essays, and letters of recommendation.
- **Stay Positive:** Remind yourself that you've prepared as well as you can, and that doing your best is what matters. Keep a positive attitude, and don't let a few difficult questions shake your confidence.

6. Use Your Breaks Wisely

The SAT includes scheduled breaks between sections. Use these breaks to recharge and prepare for the next part of the test.

- **Stretch and Move Around:** Get up, stretch your muscles, and walk around to increase blood flow and reduce physical tension.
- **Hydrate and Snack:** Drink water and have a light snack to keep your energy levels up. Avoid heavy foods that might make you feel sluggish.
- **Clear Your Mind:** Use the break to relax your mind. Don't dwell on the previous section—focus on preparing yourself mentally for the next one.

By incorporating these stress and anxiety management techniques into your test preparation and test-day routine, you can approach the SAT with confidence and stay calm under pressure. Remember, the key to managing test anxiety is preparation, mindfulness, and a positive mindset.

Building Confidence

Confidence is a crucial factor in performing well on the SAT. When you believe in your abilities and approach the test with a positive mindset, you're more likely to stay focused, manage stress, and make better decisions. Here's how to build and maintain confidence leading up to and during the SAT.

Building Confidence

1. Prepare Thoroughly

Confidence often stems from knowing that you've prepared to the best of your ability. The more you practice, the more familiar you become with the test format, question types, and time constraints, which in turn boosts your confidence.

- **Set a Study Plan:** Create a realistic study schedule that allows you to gradually build your skills over time. Consistent practice is key to feeling prepared.
- **Use Official Practice Tests:** Take full-length, timed practice tests to simulate the test-day experience. Reviewing your performance on these tests helps you identify strengths and areas for improvement, reinforcing your confidence.
- **Track Your Progress:** Keep a record of your practice test scores and note any improvements. Seeing your progress over time can be a powerful confidence booster.

2. Focus on Your Strengths

While it's important to work on areas that need improvement, don't forget to recognize and build on your strengths. Knowing that you have mastered certain parts of the test can give you a confidence boost.

- **Identify Your Strong Areas:** Take note of the sections or question types where you consistently perform well. Remind yourself of these strengths as you approach the test.
- **Leverage Your Strengths on Test Day:** Start with the questions or sections you're most comfortable with. This approach can help you build momentum and confidence as you move through the test.

3. Develop a Positive Mindset

Your mindset plays a significant role in your test performance. Cultivating a positive attitude can help you manage stress and stay focused during the exam.

- **Practice Positive Self-Talk:** Replace negative thoughts with positive affirmations. Instead of thinking, "I'm not good at this," tell yourself, "I've practiced this, and I can do it."
- **Visualize Success:** Spend a few minutes each day visualizing yourself successfully completing the SAT. Picture yourself feeling calm, confident, and in control as you answer each question.

- **Embrace Challenges:** View difficult questions as opportunities to demonstrate your skills, rather than obstacles. Remind yourself that you've faced and overcome challenges in your preparation.

4. Simulate Test Day Conditions

Familiarity with the test environment and conditions can significantly reduce anxiety and boost confidence. By simulating test day conditions during your practice sessions, you'll be better prepared for the real thing.

- **Take Practice Tests Under Realistic Conditions:** Complete practice tests in a quiet environment with no distractions. Use a timer to simulate the actual test timing, and follow the same break schedule as the real SAT.
- **Wear Comfortable Clothing:** Dress as you would on test day to ensure you're comfortable during the exam. This may seem minor, but physical comfort can impact your ability to focus.
- **Use Official Materials:** Practice with materials that closely resemble the actual test, such as official SAT practice tests and questions. Familiarity with the test format will help you feel more confident.

5. Manage Pre-Test Nerves

Feeling nervous before a big test is normal, but it's important to manage these nerves so they don't interfere with your performance.

- **Get Plenty of Rest:** Make sure you get enough sleep in the days leading up to the test, especially the night before. Being well-rested helps you stay sharp and focused.
- **Eat a Balanced Breakfast:** On test day, eat a nutritious breakfast that will sustain your energy levels throughout the morning. Avoid heavy or sugary foods that might make you feel sluggish.
- **Arrive Early:** Plan to arrive at the test center early so you have time to settle in and get comfortable. Rushing at the last minute can increase anxiety, so give yourself plenty of time.

6. Have a Test-Day Strategy

Going into the test with a clear plan can help you feel more in control and confident.

- **Start with Easy Questions:** Begin with the questions you find easiest. This helps you build momentum and secure points early on.
- **Use Your Breaks:** Take advantage of the breaks between sections to stretch, hydrate, and reset your focus. This will help you maintain your energy and concentration throughout the test.
- **Keep an Eye on the Clock:** Be mindful of the time, but don't let it stress you out. Pacing yourself effectively ensures you have time to answer as many questions as possible.

7. Reflect on Past Successes

Remind yourself of past academic successes and challenges you've overcome. Reflecting on these achievements can boost your confidence and reinforce the belief that you can succeed on the SAT.

- **Recall Difficult Tests:** Think about other challenging tests or exams you've taken in the past. Remember how you prepared, how you felt going in, and how you ultimately succeeded.
- **Celebrate Small Wins:** Acknowledge the progress you've made during your SAT preparation. Each practice test, study session, and improvement is a step toward achieving your goal.

Building confidence for the SAT is about preparation, mindset, and self-belief. By following these strategies, you'll be better equipped to approach the test with a calm and focused attitude, ready to perform your best.

Simulated Test Environments

Creating a simulated test environment is one of the most effective ways to prepare for the SAT. By replicating the conditions you'll experience on test day, you can reduce anxiety, improve your time management skills, and become more comfortable with the test format. Here's how to set up and practice in a simulated test environment.

Simulated Test Environments

1. Recreate the Test Setting

To make your practice sessions as effective as possible, try to mimic the actual test environment as closely as you can. This means minimizing distractions and ensuring that the conditions are similar to what you'll experience on test day.

- **Choose a Quiet Space:** Find a quiet place where you won't be disturbed. Avoid studying in a busy or noisy area where you might be distracted.
- **Sit at a Desk or Table:** Take your practice test sitting at a desk or table, just as you would on test day. This helps you get used to the posture and setting you'll experience during the actual SAT.
- **Turn Off Electronics:** Turn off your phone, computer, and any other electronic devices that could distract you. The goal is to create an environment that mirrors the focus required during the SAT.

2. Use a Timer

Timing is a crucial part of the SAT, so practicing with a timer is essential. This helps you get used to the pacing of each section and ensures you can complete the test within the allotted time.

- **Use an Accurate Timer:** Set a timer for each section of the SAT according to the official time limits (e.g., 65 minutes for Reading, 35 minutes for Writing and Language, 80 minutes for Math). Stick to these limits to simulate the real experience.
- **Practice Time Management:** Work on pacing yourself during the test. Aim to finish each section with a few minutes to spare so you can review your answers. Practicing with a timer helps you become more aware of how much time you have left and adjust your pace accordingly.

3. Follow the SAT Structure

Practice the entire SAT in one sitting to build your stamina and get used to the test's length. Taking the test in sections is useful for study purposes, but full-length practice tests are vital for simulating the real experience.

- **Stick to the Order:** Take the sections in the same order they'll appear on the SAT: Reading, Writing and Language, Math (No Calculator), Math (Calculator), and Essay (if applicable).
- **Take Scheduled Breaks:** The SAT includes scheduled breaks, so take these breaks during your practice test as well. Use this time to stretch, hydrate, and reset your focus, just as you would on test day.

4. Use Official Practice Tests

The best way to simulate the SAT is by using official practice tests provided by the College Board. These tests are designed to reflect the content and format of the actual SAT, making them the most reliable resources for preparation.

- **Download Official Tests:** The College Board offers free official SAT practice tests on its website. These tests are exact replicas of the real exam, including the types of questions and the timing for each section.
- **Print the Tests:** If possible, print the practice tests and take them on paper. This will more closely resemble the actual test-taking experience, where you'll be working with a paper booklet and a separate answer sheet.

5. Review Your Practice Test

After completing a simulated test, it's important to review your performance carefully. This helps you identify areas where you need to improve and adjust your study plan accordingly.

- **Check Your Answers:** Go through each section and compare your answers to the correct ones provided in the answer key. Note where you made mistakes and try to understand why.
- **Analyze Your Timing:** Review how well you managed your time in each section. Did you run out of time in any part? Did you have time left over? Use this analysis to adjust your pacing strategy for the next practice test.
- **Identify Patterns:** Look for patterns in the types of questions you got wrong. Are there specific content areas or question types that you consistently struggle with? Focus your study efforts on these areas.

6. Simulate Test-Day Conditions

In addition to practicing under timed conditions, it's helpful to simulate other aspects of test day to reduce anxiety and improve comfort.

- **Eat a Test-Day Breakfast:** On the morning of your practice test, eat the same type of breakfast you plan to have on test day. This helps you determine what foods give you sustained energy without making you feel sluggish.
- **Dress Comfortably:** Wear the clothes you plan to wear on test day, including layers in case the testing room is too cold or too warm. Being comfortable physically can help you focus mentally.
- **Practice Using Test Materials:** If you plan to use a specific calculator or set of pencils on test day, use them during your practice test as well. Familiarity with your materials reduces the chances of test-day surprises.

7. Reflect on Your Experience

After each simulated test, take some time to reflect on how you felt during the test and what you can improve for next time.

- **Note Your Comfort Level:** Were there moments when you felt particularly anxious or distracted? Consider what triggered these feelings and how you might address them.
- **Plan for Improvement:** Based on your reflections, make a plan for your next practice session. Set specific goals, such as improving your pacing in the Math section or staying focused during the Reading section.

By practicing in a simulated test environment, you can reduce test-day anxiety, improve your time management skills, and become more familiar with the SAT format. This preparation will help you feel more confident and comfortable when the real test day arrives.

Day of the Test

The day of the SAT is finally here, and it's important to approach it with a clear plan to ensure everything goes smoothly. In this section, we'll cover the key things you need to do on test day, including what to bring, how to stay calm, and last-minute tips for success.

What to Bring

Being well-prepared with the right materials can help you feel more confident and reduce stress on test day. Here's a checklist of essential items to bring with you:

1. Admission Ticket

- Print your SAT admission ticket from the College Board website and bring it with you to the test center. Without it, you may not be allowed to take the test.

2. Valid Photo ID

- Bring a government-issued or school-issued photo ID, such as a driver's license, passport, or school ID card. Make sure the ID is current and matches the name on your admission ticket.

3. #2 Pencils and Erasers

- Bring at least two or three #2 pencils with good erasers. Mechanical pencils and pens are not allowed, so stick to standard pencils.

4. Approved Calculator

- Bring an approved calculator for the Math sections. Make sure it's fully charged or has fresh batteries. Familiarize yourself with its functions before test day.

5. Extra Batteries

- If your calculator runs on batteries, bring an extra set just in case. While you can't change batteries during a section, you can do so during breaks.

6. A Watch (without Alarms)

- A simple watch can help you keep track of time during the test, especially since some test centers may not have a visible clock. Make sure the watch doesn't have an audible alarm, as this could disrupt the test.

7. Snacks and Water

- Bring a small, healthy snack and a bottle of water for the breaks. Choose snacks that provide sustained energy, like nuts, fruit, or granola bars.

8. Comfortable Clothing

- Dress in layers so you can adjust to the temperature of the test room. You want to be as comfortable as possible during the exam.

9. Face Mask and Hand Sanitizer

- Depending on current health guidelines, you may be required to wear a face mask during the test. Bring a mask and a small bottle of hand sanitizer for hygiene.

Test Day Checklist

A checklist can help you stay organized and ensure that you don't forget anything important. Here's a simple test day checklist to follow:

Night Before the Test:

- Pack your bag with all the items from the "What to Bring" list.
- Set your alarm early to allow plenty of time to get ready and arrive at the test center.
- Review your admission ticket and ID to make sure they're correct.
- Plan your route to the test center, including backup options in case of unexpected delays.
- Get a good night's sleep—aim for at least 8 hours.

Morning of the Test:

- Eat a balanced breakfast that includes protein, healthy fats, and complex carbohydrates.
- Double-check that you have your admission ticket, ID, pencils, calculator, and other essentials.
- Leave for the test center with plenty of time to account for traffic or other delays.
- Stay calm and focused—remind yourself of the preparation you've done and your strategies for the test.

At the Test Center:

- Arrive early and locate your testing room.
- Follow all instructions from test center staff and be respectful of other test-takers.
- Use the bathroom before the test begins to avoid distractions during the exam.
- Settle into your seat, arrange your materials, and take a few deep breaths to center yourself.

Last-Minute Tips

As the test begins, it's important to stay focused and use the strategies you've practiced. Here are some last-minute tips to keep in mind:

1. Stay Calm and Collected

- Remember that nerves are normal, but don't let them overwhelm you. Take deep breaths and remind yourself that you're prepared.

2. Read Instructions Carefully

- Even if you're familiar with the test format, take a moment to read the instructions for each section. This will help you avoid careless mistakes.

3. Pace Yourself

- Keep an eye on the clock, but don't rush. Answer the questions you find easiest first, and come back to the more difficult ones if you have time.

4. Double-Check Your Answers

- If you finish a section early, use the extra time to review your answers. Look for any questions you might have skipped or answered too quickly.

5. Stay Positive

- If you encounter a challenging question, don't panic. Make an educated guess, mark it if necessary, and move on. Stay focused on the next question and keep a positive mindset.

6. Use Breaks Wisely

- During breaks, stretch, drink water, and have a light snack. Clear your mind and prepare yourself for the next section.

7. Trust Your Preparation

- You've spent time studying and practicing, so trust in your preparation. Confidence in your abilities can help you perform better and reduce stress.

Section 5: Motivational and Inspirational Resources

Preparing for the SAT can be a challenging journey, but with the right mindset and inspiration, you can stay motivated and focused on your goals. This section provides you with success stories, motivational quotes, and practical tips to keep your spirits high as you work toward achieving your best score.

Success Stories

Hearing from others who have successfully navigated the SAT can be incredibly motivating. These real-life stories demonstrate that with dedication, hard work, and the right strategies, you can overcome challenges and achieve your goals.

Story 1: Overcoming Test Anxiety

- **Student:** Emma had always struggled with test anxiety, particularly during standardized exams. The pressure to perform well on the SAT made her anxious, leading to restless nights and difficulty focusing during practice tests.
- **Strategy:** Emma focused on building her confidence by practicing mindfulness techniques and deep breathing exercises. She also created a consistent study routine that included regular breaks and stress-relief activities like yoga.
- **Outcome:** On test day, Emma used her mindfulness techniques to stay calm and centered. She paced herself well, managed her anxiety, and ultimately scored higher than she expected. Her SAT score helped her gain admission to her top-choice college.

Story 2: From Struggling in Math to Mastery

- **Student:** Carlos had always found math challenging, particularly algebra and geometry. He often felt frustrated and discouraged when he couldn't grasp certain concepts, leading to poor performance on math-related practice tests.
- **Strategy:** Carlos decided to tackle his weaknesses head-on by dedicating extra time to math. He sought help from a tutor, used online resources, and practiced problems daily until he started to see improvement.
- **Outcome:** Through consistent practice and support, Carlos transformed his approach to math. On test day, he felt confident in his abilities and performed significantly better in the math sections than he had during his initial practice tests. His score improvement was a key factor in his acceptance to a competitive engineering program.

Story 3: Balancing SAT Prep with a Busy Schedule

- **Student:** Priya was a high-achieving student with a packed schedule that included extracurricular activities, part-time work, and a demanding course load. Finding time to study for the SAT felt overwhelming, and she struggled to fit in effective prep sessions.

- **Strategy:** Priya created a detailed study schedule that integrated SAT prep into her daily routine, even if it was just 30 minutes a day. She used short, focused study sessions during breaks or after school and maximized her weekends for full-length practice tests.
- **Outcome:** By staying organized and making the most of her limited time, Priya was able to prepare effectively for the SAT without sacrificing her other commitments. Her disciplined approach paid off, and she achieved a score that exceeded her expectations, earning her a scholarship to her dream college.

These success stories show that no matter your starting point, with persistence and the right strategies, you can achieve your SAT goals. Use these stories as inspiration to keep pushing forward, even when the preparation process feels challenging.

Inspirational Quotes and Tips

Sometimes, a few words of wisdom can provide the motivation you need to keep going. This section offers a collection of inspirational quotes and practical tips to help you stay focused and energized as you prepare for the SAT.

Inspirational Quotes

1. **"Success is the sum of small efforts, repeated day in and day out." — Robert Collier**
   - **Tip:** Consistency is key when preparing for the SAT. Even small, daily study sessions can add up to significant progress over time. Focus on making steady, incremental improvements.
2. **"The only way to do great work is to love what you do." — Steve Jobs**
   - **Tip:** While studying for the SAT might not always be fun, try to find aspects of the process that you enjoy. Whether it's mastering a challenging math problem or improving your reading speed, celebrate the small victories along the way.
3. **"Don't watch the clock; do what it does. Keep going." — Sam Levenson**
   - **Tip:** Time management is crucial during the SAT, but don't let the ticking clock distract you. Focus on one question at a time, and keep moving forward. Trust in the strategies you've practiced.
4. **"Believe you can and you're halfway there." — Theodore Roosevelt**
   - **Tip:** Confidence plays a huge role in test-taking success. Believe in your abilities and the preparation you've put in. A positive mindset can help you stay calm and perform your best on test day.
5. **"You are never too old to set another goal or to dream a new dream." — C.S. Lewis**
   - **Tip:** Whether it's setting a new study goal or aspiring to achieve a higher score, remember that it's never too late to improve. Keep pushing yourself to reach new heights.
6. **"The harder you work for something, the greater you'll feel when you achieve it." — Unknown**
   - **Tip:** Preparing for the SAT takes hard work and dedication, but the sense of accomplishment you'll feel when you reach your goal makes it all worthwhile. Stay focused on the end result.
7. **"It always seems impossible until it's done." — Nelson Mandela**
   - **Tip:** The SAT might feel overwhelming at first, but remember that every student who has succeeded on the test started where you are now. Take it one step at a time, and soon you'll look back at how far you've come.

Practical Tips

1. Break Down Your Goals

- Instead of focusing on your overall SAT score, break it down into smaller, more manageable goals. Aim to improve your score in specific sections or question types. This approach makes the task less daunting and allows you to celebrate progress along the way.

2. Create a Positive Study Environment

- Set up a study space that is comfortable, well-lit, and free of distractions. Surround yourself with motivational quotes, music, or anything else that keeps you focused and inspired. A positive environment can boost your productivity and mood.

3. Stay Organized with a Study Plan

- Develop a detailed study plan that outlines what you need to accomplish each day or week. Having a clear plan helps you stay on track and ensures you're covering all the material you need to review before the test.

4. Reward Yourself

- Give yourself small rewards for reaching study milestones. Whether it's a break, a treat, or time to relax, rewarding yourself can keep you motivated and prevent burnout.

5. Visualize Success

- Spend a few minutes each day visualizing yourself successfully completing the SAT. Imagine feeling confident, answering questions correctly, and achieving your goal score. Visualization can reinforce a positive mindset and help reduce anxiety.

6. Don't Be Afraid to Ask for Help

- If you're struggling with certain topics or question types, seek help from teachers, tutors, or online resources. Getting the support you need can make a big difference in your preparation and confidence.

7. Focus on Progress, Not Perfection

- It's easy to get discouraged if you don't see immediate improvement, but remember that progress is more important than perfection. Every practice session, no matter how small, brings you closer to your goal.

Use these quotes and tips as a source of motivation throughout your SAT preparation. When the journey feels tough, remind yourself of why you're working hard and keep your eyes on the prize.

Goal Setting and Achievement

Setting clear, realistic goals is essential to staying motivated and on track during your SAT preparation. This section will guide you through the process of setting effective goals, tracking your progress, and celebrating your successes.

Setting Realistic Goals

1. Start with a Clear Vision

Before setting specific goals, take a moment to reflect on what you want to achieve with your SAT score. Your overall vision might include getting into a particular college, qualifying for scholarships, or simply achieving a personal best.

- **Define Your Purpose:** Understand why you're taking the SAT and what you hope to accomplish. Having a clear purpose will keep you motivated when the preparation process gets challenging.
- **Example:** "I want to score 1400 on the SAT to increase my chances of being admitted to my top-choice college."

2. Break Down Your Vision into Specific Goals

Once you have a clear vision, break it down into smaller, specific goals. These should be concrete, measurable, and time-bound to help you stay focused and motivated.

- **Set Section-Specific Goals:** Identify target scores for each section of the SAT. This allows you to focus on improving specific areas where you might need the most work.
    - **Example:** "I want to improve my Math score by 100 points within the next two months."
- **Create Short-Term Milestones:** Break your preparation down into weekly or monthly goals. This helps you track your progress and ensures you're steadily working toward your overall target.
    - **Example:** "This week, I will complete two practice Reading sections and review the explanations for any questions I got wrong."

3. Ensure Your Goals Are Realistic

While it's important to aim high, your goals should also be realistic and achievable given your current skills, available time, and resources.

- **Assess Your Starting Point:** Take a diagnostic test to determine your current score and identify areas for improvement. Use this as a baseline to set achievable goals.
    - **Example:** "Based on my diagnostic test, I'll aim to improve my overall score by 150 points over the next three months."
- **Consider Your Schedule:** Be realistic about the amount of time you can dedicate to studying each week. If you have a busy schedule, it's better to set smaller, more manageable goals than to overextend yourself.

    - **Example:** "With my current school and work schedule, I can realistically dedicate four hours per week to SAT prep."

Tracking Progress

1. Keep a Study Journal

A study journal is a powerful tool for tracking your progress, reflecting on your performance, and staying accountable to your goals.

- **Record Your Study Sessions:** Note down what you studied, how long you studied, and any challenges you faced. This helps you identify patterns and areas where you might need to adjust your approach.
    - **Example:** "Today, I completed a timed Writing section and struggled with punctuation questions. I'll review these types of questions in my next study session."
- **Track Practice Test Scores:** Regularly take full-length practice tests and record your scores in each section. Use these scores to monitor your progress toward your target.
    - **Example:** "My Reading score improved by 50 points since my last practice test. I'm now closer to my goal of scoring 700 in this section."

2. Use Progress Tracking Tools

There are various tools and apps available that can help you track your SAT prep progress more systematically.

- **SAT Prep Apps:** Some apps allow you to set goals, track your study time, and log your practice test scores. These can provide visual representations of your progress over time.
- **Spreadsheets:** Create a simple spreadsheet to log your practice test scores, study hours, and completed sections. Spreadsheets can help you see your progress at a glance and adjust your study plan accordingly.

3. Reflect on Your Progress Regularly

Regular reflection is key to understanding how well you're progressing toward your goals and where you might need to make adjustments.

- **Weekly Check-Ins:** Set aside time each week to review your study journal or progress tracking tools. Reflect on what went well, what challenges you encountered, and how you can improve in the upcoming week.
    - **Example:** "This week, I improved my timing in the Math section but still struggled with geometry questions. Next week, I'll focus on reviewing geometry concepts."
- **Celebrate Small Wins:** Recognize and celebrate your achievements, no matter how small. This positive reinforcement will keep you motivated and committed to your goals.
    - **Example:** "I successfully completed all my study sessions this week and improved my score on practice questions. I'll treat myself to a movie night as a reward."

Celebrating Successes

1. Acknowledge Your Achievements

Every step forward in your SAT preparation is a success. Take time to acknowledge these achievements, whether it's mastering a difficult concept, improving your practice test score, or simply sticking to your study plan.

- **Reflect on Your Journey:** Consider how far you've come since you started preparing for the SAT. Reflecting on your progress can help you appreciate your hard work and dedication.
- **Share Your Successes:** Share your achievements with friends, family, or study partners. Celebrating with others can boost your confidence and provide additional motivation.

2. Reward Yourself

Rewards are a great way to stay motivated and reinforce positive study habits. Choose rewards that are meaningful to you and that you'll look forward to.

- **Set Up a Reward System:** Create a system where you reward yourself after reaching specific milestones. The rewards don't have to be extravagant—small treats can be just as motivating.
  - **Example:** "If I complete my study plan for the month and achieve my target score on the next practice test, I'll reward myself with a weekend trip."
- **Enjoy the Reward:** Once you've earned a reward, take the time to enjoy it without guilt. You've worked hard, and you deserve to celebrate your success.

3. Use Success as Motivation for Further Improvement

Each success you achieve can serve as motivation to keep pushing forward. Use your accomplishments as a reminder of what you're capable of and as fuel to reach your ultimate goals.

- **Set New Goals:** Once you've achieved a goal, set a new one that builds on your progress. Continuously striving for improvement will keep you engaged and motivated.
  - **Example:** "Now that I've improved my Math score by 100 points, I'll focus on raising my Reading score by 50 points in the next month."
- **Stay Positive:** Even if you encounter setbacks, focus on your successes and use them as a foundation to build on. Remember that progress isn't always linear, but every effort counts toward your ultimate achievement.

Setting and achieving your SAT goals is a rewarding journey that requires dedication, perseverance, and a positive mindset. By following these strategies, you can stay motivated, track your progress effectively, and celebrate your successes along the way.

Section 6: Supplementary Materials

This section provides additional resources designed to support and enhance your SAT preparation. These supplementary materials include detailed study schedules, a glossary of key terms, and an index for quick reference. The goal is to help you organize your study plan, deepen your understanding of important concepts, and easily navigate through the content of this book.

Study Schedules

Having a structured study plan is essential for effective SAT preparation. In this section, we provide you with sample study schedules that can be adapted to fit your timeline and learning style. Whether you have three months or six months to prepare, these templates will help you stay on track and cover all the necessary material.

Using the Study Schedules

Before diving into the specific schedules, it's important to understand how to use these plans effectively:

- **Set Clear Goals:** Start by identifying your target SAT score and the specific areas where you need improvement. Use this information to customize the study plan to your needs.
- **Be Realistic:** Consider your current schedule, including school, extracurricular activities, and personal commitments. Make sure the study plan you choose is realistic and sustainable given your other responsibilities.
- **Track Your Progress:** Regularly check your progress against the plan. If you find that you're falling behind, reassess your study time or the difficulty of the material. Adjustments might be necessary to stay on course.
- **Stay Flexible:** Life can be unpredictable, so it's important to remain flexible. If you need to adjust the plan due to unforeseen circumstances, do so without feeling discouraged. The key is consistency, not perfection.
- **Incorporate Breaks:** Ensure that your schedule includes breaks to prevent burnout. Short, frequent breaks can help maintain focus and retention.

3-Month Study Plan: In-Depth Guide

This 3-Month Study Plan is designed for students who need a structured approach to SAT preparation. It assumes you can dedicate 10-15 hours per week to studying. The plan is broken down into weekly goals, covering each section of the SAT, with a focus on building foundational skills, intensive practice, and test simulation.

Month 1: Foundation Building

Week 1-2: Diagnostic Test & Goal Setting

- **Goal:** Establish a baseline and identify areas for improvement.
- **Tasks:**
  1. **Take a Full-Length Diagnostic Test:** Set aside a quiet, uninterrupted 4-hour block to simulate test conditions. Use an official SAT practice test for the most accurate assessment.
     - **Action:** After completing the test, carefully review your results. Identify which sections you performed well in and which ones need improvement.
  2. **Analyze Your Results:** Break down your scores by section—Math, Evidence-Based Reading, and Writing. Look at the types of questions you missed and note any patterns.
     - **Action:** Use this analysis to set specific, measurable goals for each section. For example, "Increase Math score by 100 points" or "Improve reading comprehension speed."
  3. **Create a Study Journal:** Document your diagnostic test results, goals, and initial thoughts in a study journal. This will help you track your progress and stay motivated.
     - **Action:** Outline your goals for the next three months, including both overall score targets and specific skills you want to develop.

Week 3-4: Math Fundamentals

- **Goal:** Build a strong foundation in key math concepts.
- **Tasks:**
  1. **Focus on Heart of Algebra:** Review linear equations, inequalities, and systems of equations. These are essential topics that appear frequently on the SAT.
     - **Action:** Spend 3-4 hours each week practicing algebra problems. Use SAT prep books, online resources, or math apps that offer targeted practice.
  2. **Master Problem Solving and Data Analysis:** This includes working with ratios, percentages, and interpreting data from tables, graphs, and charts.
     - **Action:** Allocate 2-3 hours each week to practice these skills. Focus on understanding how to approach word problems and data interpretation questions.
  3. **Review Key Math Formulas:** Make sure you know the essential formulas by heart, including those for area, volume, and the Pythagorean theorem.
     - **Action:** Create flashcards or a formula sheet for quick daily review. Spend 15-20 minutes each day going over these formulas.
  4. **Timed Practice Sessions:** Start getting used to the time constraints of the SAT. Practice answering math questions under timed conditions.
     - **Action:** Set a timer for 20 minutes and see how many questions you can accurately complete. Gradually increase the time as you progress.

Month 2: Intensive Practice

Week 5-6: Evidence-Based Reading

- **Goal:** Develop critical reading skills and improve comprehension.
- **Tasks:**
    1. **Focus on Reading Comprehension Strategies:** Work on identifying main ideas, understanding the author's tone, and recognizing supporting details in passages.
        - **Action:** Spend 4-5 hours each week reading passages and answering questions. Focus on different types of passages, including historical documents, science articles, and literature.
    2. **Practice Passage-Based Questions:** Practice answering questions about paired passages and those that require interpretation of evidence.
        - **Action:** Allocate 3 hours each week to these specific types of questions. Use practice tests or reading sections from SAT prep books.
    3. **Vocabulary in Context:** Expand your vocabulary by focusing on words commonly found in SAT reading passages. Understanding words in context is crucial for answering vocabulary questions.
        - **Action:** Spend 1-2 hours each week on vocabulary building. Use flashcards, apps, or vocabulary lists specific to the SAT.
    4. **Timed Reading Practice:** Work on your reading speed without sacrificing comprehension.
        - **Action:** Set a timer and practice reading passages within the time limits of the SAT. Aim to complete a full reading section (5 passages) within 65 minutes by the end of the week.

Week 7-8: Writing and Language

- **Goal:** Improve grammar, usage, and writing skills.
- **Tasks:**
    1. **Review Grammar Rules:** Focus on the most tested grammar concepts on the SAT, including subject-verb agreement, pronoun clarity, verb tense consistency, and parallelism.
        - **Action:** Dedicate 3-4 hours each week to reviewing these rules. Use SAT prep books that offer clear explanations and plenty of practice questions.
    2. **Practice Sentence Structure and Punctuation:** Work on understanding sentence fragments, run-on sentences, comma usage, and semicolons.
        - **Action:** Allocate 2-3 hours each week to practice these concepts. Do targeted drills that focus on correcting sentences and punctuation errors.
    3. **Improving Paragraphs:** Learn how to make paragraphs clearer, more concise, and better organized. This includes revising sentences and paragraphs to improve logic and coherence.
        - **Action:** Spend 2 hours each week practicing paragraph improvement questions. These often involve rearranging sentences or choosing the best sentence to add or delete.
    4. **Timed Writing Practice:** Like the reading section, practice completing the Writing and Language section within the SAT's time constraints (35 minutes).
        - **Action:** Set a timer and complete practice sections, focusing on accuracy and speed. Aim to complete a full writing section within the allotted time by the end of the week.

Month 3: Refinement & Test Simulation

Week 9-10: Advanced Math Topics

- **Goal:** Master complex math topics that appear on the SAT.
- **Tasks:**
    1. **Focus on Passport to Advanced Math:** This includes working with quadratic equations, functions, and polynomials.
        - **Action:** Spend 3-4 hours each week reviewing these topics. Practice solving equations, graphing functions, and factoring polynomials.
    2. **Review Additional Topics in Math:** This includes geometry, trigonometry, and complex numbers. These topics are less common but can still appear on the SAT.
        - **Action:** Allocate 2-3 hours each week to practicing these advanced topics. Focus on understanding the underlying concepts rather than memorizing formulas.
    3. **Mixed Practice Sessions:** Combine all the math topics you've studied so far. Work on questions that integrate different areas of math, as this is how they will appear on the SAT.
        - **Action:** Set aside 2 hours each week for mixed practice sessions. Use full-length practice tests or custom question sets that include a variety of math topics.
    4. **Timed Practice with Calculator:** Practice using your calculator efficiently. Although not all math sections allow calculators, knowing how to use one effectively can save time.
        - **Action:** Spend 1-2 hours practicing with the calculator. Focus on problems that are easier to solve with technology, such as those involving complex arithmetic or graphing.

Week 11-12: Full-Length Practice Tests

- **Goal:** Simulate the SAT experience and refine your test-taking strategies.
- **Tasks:**
    1. **Take Full-Length, Timed Practice Tests:** Set aside a 4-hour block for each test to simulate the actual SAT experience. This includes all sections—Reading, Writing, Math (No Calculator), and Math (Calculator).
        - **Action:** Complete two full-length practice tests over the next two weeks. Treat each test as if it were the real exam—time yourself strictly, follow all test rules, and minimize distractions.
    2. **Review Your Results:** After each test, take the time to carefully review your answers. Focus on understanding why you got certain questions wrong and how you can avoid similar mistakes in the future.
        - **Action:** Spend several hours analyzing your test results. Identify patterns in your errors, whether they're due to timing issues, misunderstandings, or careless mistakes.
    3. **Refine Your Strategies:** Based on your practice test performance, make any necessary adjustments to your strategies. This might include changing your approach to certain question types or adjusting your timing.
        - **Action:** Use the final week to fine-tune your strategies. Focus on your weakest areas and practice under timed conditions to build your confidence.
    4. **Relax and Prepare Mentally:** The last few days before the SAT should focus on relaxation and mental preparation. Avoid cramming and focus on staying calm and confident.
        - **Action:** Engage in light review sessions, practice deep breathing exercises, and ensure you get plenty of rest before the actual test day.

Final Tips:

- **Stay Consistent:** Consistency is key to success. Stick to your study plan as closely as possible, and make up for any missed sessions to keep on track.
- **Seek Help When Needed:** Don't hesitate to seek help if you're struggling with certain topics. Tutors, teachers, and online resources can provide valuable support.
- **Reward Yourself:** Acknowledge your progress and reward yourself for sticking to the plan. Small rewards can keep you motivated throughout your study journey.

By following this in-depth 3-Month Study Plan, you'll be well-prepared to tackle the SAT with confidence. The plan is designed to build your skills progressively, ensuring you cover all necessary content while also practicing essential test-taking strategies.

6-Month Study Plan: In-Depth Guide

This 6-Month Study Plan is designed for students who have a longer timeline to prepare for the SAT. It allows for a more gradual and thorough review of all test sections, with ample time for practice, reflection, and adjustment. The plan assumes a commitment of 5-10 hours per week, making it manageable alongside schoolwork and other responsibilities.

Month 1-2: Baseline and Basics

Week 1-4: Diagnostic Test & Goal Setting

- **Goal:** Establish a baseline and create a customized study plan.
- **Tasks:**
  1. **Take a Full-Length Diagnostic Test:** Begin by simulating the SAT experience with a full-length practice test. This will give you a clear picture of your current strengths and weaknesses.
     - **Action:** Set aside a 4-hour block to take the test under real test conditions—quiet environment, timed sections, and minimal distractions.
  2. **Analyze Your Results:** Break down your scores by section—Math, Evidence-Based Reading, and Writing. Look for patterns in the types of questions you struggled with.
     - **Action:** Identify the top three areas where you need improvement. Set specific, measurable goals for each, such as "Improve Reading score by 50 points" or "Master algebraic expressions."
  3. **Create a Study Journal:** Use this journal to document your diagnostic test results, goals, and initial thoughts. It will help you track your progress and make adjustments as needed.
     - **Action:** Outline your goals for the next six months, including both overall score targets and specific skills you want to develop.

Week 5-8: Basic Math Review

- **Goal:** Build a solid foundation in fundamental math concepts.
- **Tasks:**
  1. **Focus on Heart of Algebra:** Review linear equations, inequalities, and systems of equations. These are core topics that frequently appear on the SAT.
     - **Action:** Dedicate 2-3 hours each week to practicing algebra problems. Use SAT prep books or online resources that offer targeted practice.
  2. **Introduction to Problem Solving and Data Analysis:** Begin working with ratios, percentages, and interpreting data from tables, graphs, and charts.
     - **Action:** Spend 1-2 hours each week on these topics. Focus on understanding how to approach word problems and data interpretation questions.
  3. **Learn Key Math Formulas:** Start memorizing essential formulas, such as those for area, volume, and the Pythagorean theorem.
     - **Action:** Create flashcards or a formula sheet for daily review. Spend 10-15 minutes each day going over these formulas.

4. **Introduction to Timed Practice:** Start practicing answering math questions under timed conditions to get used to the pace of the SAT.
    - **Action:** Set a timer for 15-20 minutes and practice solving as many questions as possible within that time. Gradually increase the duration as you improve.

Month 3-4: Comprehensive Review

Week 9-12: Evidence-Based Reading

- **Goal:** Improve reading comprehension and critical reading skills.
- **Tasks:**
    1. **Deepen Reading Comprehension:** Focus on identifying main ideas, understanding the author's tone, and recognizing supporting details in passages.
        - **Action:** Dedicate 2-3 hours each week to reading passages and answering related questions. Include a variety of passage types, such as literature, social sciences, and natural sciences.
    2. **Practice Passage-Based Questions:** Work on paired passages and those that require interpretation of evidence, which are common on the SAT.
        - **Action:** Allocate 2 hours each week to practicing these specific question types. Use practice tests or reading sections from SAT prep books.
    3. **Build Vocabulary in Context:** Expand your vocabulary by focusing on words that commonly appear in SAT reading passages. Understanding words in context is crucial for success.
        - **Action:** Spend 1 hour each week on vocabulary building. Use flashcards, apps, or SAT-specific vocabulary lists.
    4. **Timed Reading Practice:** Practice completing reading passages within the SAT's time constraints to improve speed and accuracy.
        - **Action:** Set a timer for 60-65 minutes and aim to complete a full reading section (5 passages) within that time by the end of the month.

Week 13-16: Writing and Language

- **Goal:** Master grammar, usage, and effective writing skills.
- **Tasks:**
    1. **Review Essential Grammar Rules:** Focus on the most tested grammar concepts, including subject-verb agreement, pronoun clarity, verb tense consistency, and parallelism.
        - **Action:** Spend 2-3 hours each week reviewing these rules and completing related practice questions.
    2. **Practice Sentence Structure and Punctuation:** Work on understanding sentence fragments, run-on sentences, and correct punctuation usage.
        - **Action:** Allocate 1-2 hours each week to practice these concepts. Focus on exercises that involve correcting sentences and improving paragraph coherence.
    3. **Enhance Paragraph Improvement Skills:** Learn how to make paragraphs clearer and better organized, which includes revising sentences and improving logic and coherence.
        - **Action:** Spend 1-2 hours each week on paragraph improvement exercises. Practice choosing the best sentence to add or delete within a paragraph.
    4. **Timed Writing Practice:** Get used to the timing of the Writing and Language section by completing sections under timed conditions (35 minutes).
        - **Action:** Set a timer and complete a full writing section within the allotted time. Review your answers to understand any mistakes.

Month 5-6: Focused Practice & Test Simulation

Week 17-20: Advanced Math Topics

- **Goal:** Master more complex math concepts that are likely to appear on the SAT.
- **Tasks:**
    1. **Focus on Passport to Advanced Math:** This includes quadratic equations, functions, and polynomials—topics that require higher-level problem-solving skills.
        - **Action:** Dedicate 3-4 hours each week to studying these topics. Practice solving quadratic equations, graphing functions, and manipulating polynomials.
    2. **Review Additional Math Topics:** Cover geometry, trigonometry, and complex numbers. Although these topics are less common, they can still appear on the SAT.
        - **Action:** Spend 2-3 hours each week on these advanced topics. Ensure you understand the concepts, as memorization alone won't be enough.
    3. **Integrated Practice Sessions:** Combine all math topics studied so far. Work on problems that require knowledge from different areas of math.
        - **Action:** Set aside 2 hours each week for mixed practice sessions. Use custom question sets or full-length practice tests that include a variety of math topics.
    4. **Timed Calculator Practice:** Practice using your calculator efficiently, as this can save valuable time during the SAT.
        - **Action:** Spend 1-2 hours each week practicing calculator-allowed sections. Focus on problems that involve complex arithmetic or require graphing.

Week 21-24: Full-Length Practice Tests & Final Review

- **Goal:** Simulate the SAT experience and finalize test-taking strategies.
- **Tasks:**
    1. **Take Full-Length Practice Tests:** Dedicate 4-hour blocks to complete full-length practice tests under real test conditions. This will help you build endurance and refine your strategies.
        - **Action:** Complete two full-length practice tests over these weeks. Follow all test rules, including strict timing and minimal distractions.
    2. **Detailed Review of Results:** After each test, carefully review your answers. Focus on understanding why you got questions wrong and how to avoid similar mistakes.
        - **Action:** Spend several hours analyzing your performance. Identify recurring errors, whether due to timing, misunderstandings, or careless mistakes.
    3. **Refine Test-Taking Strategies:** Based on your practice test results, make necessary adjustments to your approach. This could include pacing, question prioritization, or guessing strategies.
        - **Action:** Use the remaining weeks to fine-tune your strategies. Focus on your weakest areas and practice under timed conditions to build confidence.
    4. **Final Review & Relaxation:** In the last few days before the SAT, focus on light review and mental preparation. Avoid cramming and prioritize staying calm and confident.
        - **Action:** Engage in light review sessions, practice relaxation techniques, and ensure you get plenty of sleep leading up to the test day.

Final Tips:

- **Consistency is Key:** Stick to the plan as closely as possible, but remain flexible. Adjust the schedule if necessary to accommodate unexpected events or challenges.
- **Seek Support When Needed:** If you encounter difficult topics, don't hesitate to seek help from teachers, tutors, or online resources.
- **Track Your Progress:** Regularly update your study journal to monitor your progress. This will help you stay motivated and make necessary adjustments along the way.
- **Reward Yourself:** Celebrate small victories and milestones to keep yourself motivated throughout the six-month preparation period.

By following this comprehensive 6-Month Study Plan, you'll build a strong foundation, gradually improve your skills, and refine your test-taking strategies, all while staying motivated and confident in your SAT preparation journey.

Daily Study Schedule & Customizing Your Study Journal

Daily Study Schedule

To effectively manage your time and ensure you're making consistent progress, it's important to establish a daily study schedule. This schedule should be tailored to your individual needs, preferences, and the amount of time you have available each day. Here's a sample daily study schedule to guide you:

Sample Daily Study Schedule

Morning (60-90 minutes):

- **Math Practice:**
  - **Warm-Up (10 minutes):** Start with 2-3 quick review questions from previously studied material.
  - **Focused Practice (30-45 minutes):** Work on a specific math topic, such as algebra or geometry, using practice problems or drills.
  - **Review (10-15 minutes):** Go over any mistakes made during practice. Revisit concepts that need reinforcement.

Afternoon (60-90 minutes):

- **Evidence-Based Reading & Writing:**
  - **Reading Comprehension (30-45 minutes):** Read one or two passages, then answer related questions. Focus on improving reading speed and comprehension.
  - **Grammar and Writing Practice (30-45 minutes):** Review grammar rules and practice identifying errors in sample sentences. Work on improving sentence structure and clarity.

Evening (30-60 minutes):

- **Review & Reflection:**
  - **Daily Review (15-30 minutes):** Go over what you learned during the day. Review notes, flashcards, or problem sets.
  - **Journal Entry (15-30 minutes):** Reflect on your progress, challenges, and any adjustments needed in your study plan. Note any areas that require additional focus.

Optional:

- **Extra Practice (30 minutes):** If time allows, dedicate an additional 30 minutes to working on your weakest area or taking a timed mini-quiz.

Tips for Customizing Your Daily Schedule:

- **Prioritize Your Weakest Areas:** Focus more time on sections where you need the most improvement. This could mean spending extra time on math if it's your weakest subject, or on reading comprehension if you struggle with time management.
- **Adjust Based on Your Energy Levels:** If you're more focused in the morning, start with the most challenging tasks. Reserve lighter review sessions for the afternoon or evening.
- **Include Breaks:** Don't forget to take short breaks between study sessions to rest and recharge. A 5-10 minute break can help maintain focus and prevent burnout.
- **Stay Flexible:** If you have a particularly busy day, adjust your schedule accordingly. Even a shorter study session is better than skipping it entirely.

Customizing Your Study Journal

Your study journal is a vital tool in tracking your progress, reflecting on your learning, and staying organized throughout your SAT preparation. Here's how to set up and customize your study journal:

Setting Up Your Study Journal

1. Choose the Right Format:

- **Digital vs. Paper:** Decide whether you prefer a digital journal (such as an app or a document on your computer) or a traditional paper journal. Digital journals are easily editable and portable, while paper journals offer a tangible, distraction-free way to document your journey.

2. Organize by Sections:

- **Weekly Goals:** At the start of each week, outline your study goals. Include specific tasks, such as "Complete 50 algebra problems" or "Read two SAT passages and answer questions."
- **Daily Logs:** Each day, write a brief summary of what you studied, how long you spent, and any challenges you faced. Include notes on what you learned and what still needs work.
- **Progress Tracking:** Create a section to track your scores on practice tests, quizzes, or individual sections. Use graphs or charts to visualize your improvement over time.
- **Reflection Section:** Dedicate space for weekly or monthly reflections. Write about what's working well, where you need to adjust your plan, and how you're feeling about your progress.

Customizing Your Study Journal

1. Include Motivational Elements:

- **Quotes & Inspiration:** Add motivational quotes, success stories, or personal reminders to keep yourself inspired. This can be especially helpful on challenging days.
- **Achievements & Rewards:** Keep track of milestones and rewards. For example, "Completed all reading sections for the week – reward: a movie night."

2. Use Color Coding:

- **Highlight Key Areas:** Use different colors to highlight sections that need more attention. For example, use red for areas that require immediate focus, and green for areas where you've made good progress.
- **Organize by Subject:** Assign a color to each SAT section (Math, Reading, Writing) to make your journal more organized and visually appealing.

3. Incorporate Practice Test Reviews:

- **Detailed Analysis:** After each practice test, write a detailed review in your journal. Break down your performance by section, question type, and timing.
- **Action Plan:** Based on your analysis, outline an action plan to address weaknesses before your next test. Document specific strategies you'll use to improve.

4. Adjust Your Plan as Needed:

- **Flexible Scheduling:** If you notice that certain study strategies aren't working, or if you're not making the progress you hoped for, use your journal to adjust your study plan. Document the changes and monitor their effectiveness.
- **Personal Reflection:** Include personal reflections on your mental and emotional state. Preparing for the SAT can be stressful, so use your journal to express any concerns or anxieties and brainstorm ways to manage them.

Final Tips:

- **Consistency:** Make journaling a daily habit. Even if it's just a quick note, regular entries will help you stay on track and mindful of your progress.
- **Review Regularly:** Periodically review your journal to see how far you've come and what adjustments may still be needed.
- **Be Honest:** Your journal is a personal tool, so be honest with yourself about your strengths, weaknesses, and overall progress.

By using a detailed daily study schedule and a well-organized study journal, you'll create a personalized roadmap to SAT success. These tools will not only help you manage your time effectively but also keep you motivated, focused, and on the path to achieving your best possible score.

Section 7: Full-Length Test

Introduction

Welcome to the Full-Length Test section, a crucial part of your SAT preparation journey. This section provides you with full-length practice tests that closely mirror the actual SAT exam. These tests are designed to help you simulate the experience of taking the SAT, assess your current readiness, and identify areas that require further improvement.

Why Full-Length Practice Tests Are Essential

Taking full-length practice tests is one of the most effective ways to prepare for the SAT. These tests serve several important purposes:

1. **Simulating Test Day Conditions:** The SAT is a rigorous, nearly four-hour exam that requires sustained focus and endurance. By taking practice tests under real test conditions, including timing each section and working without interruptions, you'll build the stamina needed for test day.
2. **Assessing Your Progress:** Full-length tests provide a comprehensive assessment of your current abilities. They allow you to see how well you perform across all sections—Math, Reading, and Writing—and to track your progress over time.
3. **Identifying Weaknesses:** Practice tests highlight your strengths and weaknesses. By analyzing your performance, you can pinpoint the areas where you need to focus your study efforts, whether it's mastering complex algebra problems, improving your reading speed, or refining your grammar skills.
4. **Improving Time Management:** One of the biggest challenges of the SAT is managing your time effectively. Practice tests help you develop a sense of pacing, allowing you to practice completing each section within the allotted time. This is critical to ensuring that you can answer all the questions without rushing or running out of time.
5. **Building Confidence:** Familiarity with the test format and question types builds confidence. As you complete more practice tests, you'll become more comfortable with the SAT's structure and demands, reducing test-day anxiety and boosting your overall performance.

How to Use This Section

This section contains multiple full-length practice tests, each designed to offer a unique set of challenges. Here's how to get the most out of these tests:

1. **Simulate Real Test Conditions:**
    - Find a quiet, distraction-free environment where you can complete the entire test without interruptions.
    - Use a timer to strictly adhere to the time limits for each section. This will help you replicate the pressure and pace of the actual SAT.
2. **Take the Tests Seriously:**
    - Approach each practice test as if it were the real exam. This means following all the rules, such as using only approved calculators and not taking unauthorized breaks.
3. **Review Your Answers Carefully:**
    - After completing each test, spend time thoroughly reviewing your answers. For every incorrect response, determine whether the error was due to a misunderstanding of the material, a careless mistake, or a time management issue.
4. **Analyze Your Performance:**
    - Use the detailed answer explanations provided after each test to understand why the correct answers are right and why the incorrect options are wrong. This analysis is crucial for improving your understanding and avoiding similar mistakes in the future.
5. **Track Your Progress:**
    - Keep a record of your scores for each practice test. Use this record to track your progress over time and adjust your study plan as needed. Focus on the areas where you see the most consistent challenges.
6. **Reflect and Adjust:**
    - Use your study journal to reflect on each test experience. Document what went well and what didn't, and make adjustments to your study plan accordingly. This could include dedicating more time to specific subjects, altering your test-taking strategies, or improving your time management skills.

What to Expect in This Section

Each full-length practice test in this section includes:

- **Reading Test:** A series of passages followed by multiple-choice questions designed to assess your reading comprehension and critical reading skills.
- **Writing and Language Test:** Passages that require you to identify and correct grammatical errors, improve sentence structure, and enhance the clarity and effectiveness of written communication.
- **Math Test – No Calculator:** A set of math problems that test your ability to solve problems using logic and understanding of fundamental math concepts without the aid of a calculator.
- **Math Test – Calculator:** A series of math questions where calculator use is permitted, focusing on more complex calculations and problem-solving scenarios.

The Full-Length Test section is a powerful tool in your SAT preparation arsenal. By taking these practice tests seriously and learning from each experience, you'll be well-prepared to face the real SAT with confidence and skill. Remember, consistency and dedication are key—use these tests to refine your abilities, build your test-taking endurance, and ultimately achieve your best possible SAT score.

Full-length Test

Section 1: Reading and Writing

Module 1: Reading and Writing

Total Time: 32 minutes

Total Questions: 25

Passage 1 (Literature):

Passage Title: *A Winter's Tale*

In the depths of winter, the old town nestled in the valley took on a magical quality. The houses, with their steeply pitched roofs, were blanketed in a thick layer of snow, and smoke curled lazily from the chimneys, rising into the cold, crisp air. The streets were deserted, save for the occasional footprints left by a brave soul venturing out into the frigid weather. The trees that lined the narrow lanes were coated in a shimmering layer of frost, their branches bending under the weight of the ice.

Inside one of the cozy cottages, an elderly woman sat by the fireplace, knitting a warm scarf for her grandson. The fire crackled softly, casting a golden glow across the room and warming her cold, arthritic fingers. Her eyes, though clouded with age, sparkled with the memories of winters long past—of sleigh rides down the hill, of snowball fights with her siblings, and of nights spent huddled under thick blankets, listening to the howling wind outside.

As she worked, her thoughts turned to her grandson, who would soon return from the city for the holidays. She had always looked forward to his visits, for he brought with him tales of the bustling metropolis, of the tall buildings and the bright lights that never dimmed. Yet, in her heart, she knew that the charm of the old town, with its quiet streets and timeless traditions, would always hold a special place in his heart, just as it did in hers.

The winter nights were long and cold, but they were also filled with a certain peace—a peace that came from the knowledge that, no matter how harsh the weather, the warmth of home and family would always be there to drive away the chill.

Questions:

1. **What is the meaning of the word "arduous" as used in the passage?**
   - A) Simple
   - B) Exhausting
   - C) Delightful
   - D) Mysterious
2. **Which sentence best describes the structure of the passage?**
   - A) The author presents a series of challenges, then offers solutions.
   - B) The author introduces a conflict and then resolves it.
   - C) The author compares two opposing viewpoints.
   - D) The author narrates a sequence of events.
3. **What is the tone of the passage?**
   - A) Somber and reflective
   - B) Warm and nostalgic
   - C) Detached and clinical
   - D) Joyful and celebratory
4. **What can be inferred about the relationship between the woman and her grandson?**
   - A) It is distant and formal.
   - B) It is strained by distance.
   - C) It is close and affectionate.
   - D) It is marked by misunderstanding.
5. **Which of the following best describes the theme of the passage?**
   - A) The inevitability of change and loss
   - B) The enduring value of tradition and family
   - C) The conflict between rural and urban life
   - D) The harshness of winter and its challenges

Passage 2 (History/Social Studies):

Passage Title: *The Rise of the Factory System*

In the early 19th century, the landscape of industry and labor underwent a profound transformation with the rise of the factory system. This new model of production marked a departure from the small-scale, home-based manufacturing that had been the norm for centuries. Instead of individual artisans crafting goods by hand in their homes or small workshops, factories brought together large numbers of workers under one roof, where they operated machinery to produce goods on a much larger scale.

The shift to factory-based production was driven by several factors. Technological innovations, such as the spinning jenny and the power loom, made it possible to produce textiles more quickly and efficiently than ever before. These machines required a centralized power source, such as a water wheel or a steam engine, which in turn necessitated the construction of large factories. The availability of cheap labor, particularly from women and children, also contributed to the rapid expansion of the factory system.

The factory system had far-reaching social and economic consequences. On the one hand, it led to a significant increase in productivity and economic growth. The mass production of goods lowered prices and made a wide range of products more accessible to ordinary people. On the other hand, the factory system also brought about harsh working conditions for many laborers. Factory workers often toiled for long hours in dangerous and unhealthy environments, with little regard for their well-being. Child labor was particularly widespread, with young children working in factories for meager wages under brutal conditions.

Despite these challenges, the factory system continued to expand throughout the 19th century, becoming the dominant form of industrial organization in many parts of the world. It laid the foundation for the modern industrial economy and reshaped the social fabric of societies, setting the stage for the labor movements and reforms that would follow in the decades to come.

Questions:

1. **How does the author's use of historical evidence support their argument?**
   - A) By providing statistical data
   - B) By citing authoritative sources
   - C) By recounting a historical event
   - D) By comparing past and present circumstances
2. **Which of the following best captures the tone of the passage?**
   - A) Objective and analytical
   - B) Nostalgic and sentimental
   - C) Critical and argumentative
   - D) Optimistic and hopeful

3. **What was one major consequence of the rise of the factory system?**
    - A) A decrease in the availability of goods
    - B) An improvement in working conditions
    - C) An increase in child labor
    - D) A reduction in economic productivity
4. **What can be inferred about the impact of technological innovations on the factory system?**
    - A) They played a minimal role in the development of factories.
    - B) They were the primary reason for the growth of the factory system.
    - C) They hindered the efficiency of production in factories.
    - D) They were largely unrelated to the expansion of the factory system.
5. **Which of the following best describes the author's attitude toward the factory system?**
    - A) Entirely supportive
    - B) Entirely critical
    - C) Balanced, acknowledging both positive and negative aspects
    - D) Indifferent to its effects

Passage 3 (Humanities):

Passage Title: *The Evolution of Modern Art*

The early 20th century witnessed a dramatic shift in the world of art, as artists began to move away from traditional forms and techniques, embracing new ways of seeing and representing the world around them. This period of intense experimentation and innovation gave birth to what we now call modern art—a movement that challenged established norms and redefined the boundaries of artistic expression.

One of the key features of modern art was its emphasis on abstraction. Artists like Pablo Picasso and Georges Braque pioneered the Cubist movement, which fragmented objects into geometric shapes and presented them from multiple perspectives at once. This departure from realistic representation was a radical shift, as it encouraged viewers to engage with art on a conceptual level, rather than simply admiring its technical execution.

Another significant aspect of modern art was its exploration of the subconscious mind. Surrealists like Salvador Dalí and René Magritte sought to capture the illogical and dreamlike qualities of the human psyche. Their works often featured bizarre and fantastical imagery, reflecting the influence of Freudian psychoanalysis and the belief that art could tap into deeper, unconscious truths.

In addition to these formal innovations, modern art was also deeply connected to the social and political upheavals of the time. The horrors of World War I, the rise of totalitarian regimes, and the struggles for civil rights all left their mark on the art of the era. For example, German Expressionists like Ernst Ludwig Kirchner used their work to convey the anxiety and alienation of modern life, while artists associated with the Harlem Renaissance, such as Aaron Douglas, celebrated African American culture and history in their vibrant, dynamic compositions.

As modern art evolved, it became increasingly diverse, encompassing a wide range of styles and movements, from Abstract Expressionism to Pop Art. Despite their differences, these various forms of modern art shared a common goal: to break free from the constraints of tradition and explore new possibilities for creativity and expression. This spirit of innovation continues to influence artists today, as they push the boundaries of what art can be in an ever-changing world.

Questions:

1. **What is the main idea of the passage?**
   - A) The development of modern art is largely influenced by societal changes.
   - B) Traditional art forms are losing significance in contemporary culture.
   - C) Modern art often challenges established norms and conventions.
   - D) Art critics often disagree on the value of modern artistic expressions.

2. **How did Cubist artists like Picasso and Braque challenge traditional artistic techniques?**
    - A) By using only monochromatic color schemes
    - B) By presenting objects in a fragmented, geometric form
    - C) By returning to classical styles of representation
    - D) By focusing exclusively on landscapes
3. **What influence did Freudian psychoanalysis have on Surrealist art?**
    - A) It inspired a focus on rational, logical imagery.
    - B) It encouraged the depiction of realistic, everyday scenes.
    - C) It led to the exploration of the subconscious and dreamlike imagery.
    - D) It resulted in the rejection of all traditional forms of art.
4. **According to the passage, how did the social and political climate of the early 20th century affect modern art?**
    - A) It had little impact on the themes and styles of modern artists.
    - B) It led artists to create works that reflected the turmoil of the time.
    - C) It caused a decline in the production of modern art.
    - D) It encouraged artists to focus solely on abstract forms.
5. **Which of the following best describes the shared goal of the various movements within modern art?**
    - A) To uphold traditional artistic values
    - B) To create art that was universally understood
    - C) To experiment with new forms and ideas, breaking from tradition
    - D) To focus primarily on commercial success

Passage 4 (Science):

Passage Title: *The Role of DNA in Cellular Function*

Deoxyribonucleic acid, or DNA, is often referred to as the blueprint of life. This complex molecule carries the genetic instructions necessary for the growth, development, functioning, and reproduction of all living organisms. At the core of DNA's role in cellular function is its ability to store and transmit genetic information from one generation to the next, ensuring the continuity of life.

DNA is composed of two long strands that form a double helix, a structure discovered by James Watson and Francis Crick in 1953. Each strand is made up of smaller units called nucleotides, which are arranged in a specific sequence. These sequences encode the instructions needed to build and maintain an organism. The order of the nucleotides determines the sequence of amino acids in proteins, which are the molecules responsible for most of the structure and function within a cell.

One of the critical processes involving DNA is transcription, where the information encoded in a segment of DNA is copied into a molecule of messenger RNA (mRNA). This mRNA then travels out of the cell's nucleus and into the cytoplasm, where it serves as a template for protein synthesis during a process called translation. The ribosome, a molecular machine, reads the mRNA sequence and assembles the corresponding amino acids to form a protein.

Proteins play a wide range of roles in the cell, from catalyzing chemical reactions as enzymes to providing structural support and regulating gene expression. The diversity of protein functions is directly related to the variety of genes encoded within an organism's DNA. Mutations or changes in the DNA sequence can lead to variations in protein structure and function, which can have significant consequences for the organism's health and development.

In addition to coding for proteins, DNA also has regulatory regions that control when and where genes are expressed. These regulatory elements can turn genes on or off in response to various signals, allowing cells to adapt to changes in their environment. This dynamic regulation of gene expression is crucial for processes such as cell differentiation, where cells develop into specialized types with distinct functions.

Understanding the role of DNA in cellular function has revolutionized biology and medicine. It has paved the way for advances in genetic engineering, gene therapy, and personalized medicine, where treatments can be tailored to an individual's genetic makeup. As research continues, the knowledge of DNA's role in life processes will undoubtedly lead to even more groundbreaking discoveries in the future.

Questions:

1. **What is the primary function of DNA as described in the passage?**
    - A) To provide energy for cellular processes
    - B) To store and transmit genetic information
    - C) To transport molecules within the cell
    - D) To protect the cell from external threats
2. **What is the significance of the sequence of nucleotides in DNA?**
    - A) It determines the type of cell that will be formed.
    - B) It directs the order of amino acids in proteins.
    - C) It controls the cell's ability to divide.
    - D) It ensures that the DNA remains stable over time.
3. **How does the process of transcription contribute to protein synthesis?**
    - A) By creating a copy of DNA that remains in the nucleus
    - B) By converting amino acids into proteins directly
    - C) By transferring the DNA code to mRNA for translation
    - D) By assembling proteins from existing amino acids
4. **According to the passage, what role do regulatory regions of DNA play?**
    - A) They repair damaged DNA sequences.
    - B) They facilitate the replication of DNA.
    - C) They control when and where genes are expressed.
    - D) They ensure that all cells in an organism are identical.
5. **What is one potential application of the knowledge of DNA in medicine?**
    - A) Developing treatments that target specific genetic disorders
    - B) Creating universal vaccines for all diseases
    - C) Increasing the natural lifespan of humans
    - D) Enhancing physical strength and endurance

Passage 5 (Editing):

Passage Title: *The Importance of Urban Green Spaces*

Urban green spaces, such as parks, gardens, and tree-lined streets, play a crucial role in the well-being of city dwellers. These areas not only provide a respite from the hustle and bustle of city life but also offer a range of environmental, social, and health benefits. However, maintaining and expanding urban green spaces is often overlooked in city planning, leading to a scarcity of such spaces in many densely populated areas.

Original Passage:

The benefits of urban green spaces are numerous. They improve air quality by absorbing pollutants, reduce urban heat islands by providing shade, and increase biodiversity by offering habitats for various species. Additionally, these spaces serve as social hubs where communities can gather, fostering a sense of belonging and promoting physical activity.

Questions:

1. **Which of the following sentences contains a grammatical error?**
    - A) The committee meets every Monday to discuss the agenda.
    - B) Neither the manager nor the employees are aware of the new policy.
    - C) Each of the students is required to submit their assignments on time.
    - D) The report, along with the supporting documents, was submitted yesterday.
2. **Choose the correct form of the verb to complete the sentence:**
    - "The number of participants who ______________ the event has increased significantly."
    - A) attend
    - B) attends
    - C) attending
    - D) attended
3. **Where should the comma be placed in the following sentence?**
    - "The project will require time, resources, and careful planning."
    - A) After "require"
    - B) After "time"
    - C) After "resources"
    - D) No change is necessary
4. **Which of the following sentences is punctuated correctly?**
    - A) The conference starts at 9:00 AM, however, registration begins at 8:30 AM.
    - B) The conference starts at 9:00 AM; however, registration begins at 8:30 AM.
    - C) The conference starts at 9:00 AM, however; registration begins at 8:30 AM.
    - D) The conference starts at 9:00 AM; however registration begins at 8:30 AM.

5. **In the sentence, "Furthermore urban green spaces play a key role in mitigating the effects of climate change, they absorb carbon dioxide and help lower temperatures in urban areas," which punctuation change improves the sentence?**
   - A) Insert a comma after "Furthermore"
   - B) Replace the comma after "climate change" with a semicolon
   - C) Insert a colon after "Furthermore"
   - D) Replace the comma after "spaces" with a period

Module 2: Reading and Writing

Total Time: 32 minutes

Total Questions: 27

Passage 1 (Literature):

Passage Title: *An Unexpected Visitor*

The small, isolated village of Tuckerton was known for its quiet charm and its distance from the hustle and bustle of city life. The villagers, though friendly, were wary of strangers, and rarely did anyone from outside pass through the narrow, winding roads that led to their homes. So, when a stranger appeared one cold, misty morning, it caused quite a stir.

The man was tall, with a worn coat that had seen better days. His face was obscured by a wide-brimmed hat, and his steps were slow, as if he had been walking for a long time. He made his way to the village square, where he paused to look around, taking in the unfamiliar surroundings. The villagers watched from their windows and doorways, whispering among themselves. Who was this man? What did he want? And why had he come to Tuckerton, of all places?

Despite their curiosity, no one approached the stranger. Instead, they continued to observe him from a distance, speculating about his origins and his purpose. The stranger, seemingly unfazed by the attention, walked over to the village inn and pushed open the door. Inside, the innkeeper looked up from his book, startled by the unexpected visitor. The stranger removed his hat and greeted the innkeeper with a quiet, respectful nod.

"I need a room for the night," the man said, his voice low but clear.

The innkeeper hesitated, unsure of how to respond. Tuckerton was not a place for travelers, and the inn had not had a guest in many months. But the stranger's demeanor was polite, and there was something in his eyes—a weariness, perhaps—that moved the innkeeper to offer him a room.

As the stranger settled into his quarters, the villagers continued to speculate about his identity and his reasons for visiting their remote village. Some thought he might be a lost traveler; others believed he was hiding from something—or someone. But no one could be certain, and the mystery of the stranger in Tuckerton lingered in the cold, misty air.

Questions:

1. **What can be inferred about the village of Tuckerton based on the passage?**
    - A) It is a bustling, vibrant place full of activity.
    - B) It is isolated and rarely visited by outsiders.
    - C) It is known for its hospitality and frequent visitors.
    - D) It is a dangerous place that travelers avoid.
2. **How do the villagers react to the arrival of the stranger?**
    - A) They warmly welcome him and offer assistance.
    - B) They ignore him and go about their daily routines.
    - C) They are curious and watch him cautiously from a distance.
    - D) They confront him and demand to know his intentions.
3. **What is the tone of the passage?**
    - A) Lighthearted and humorous
    - B) Suspenseful and mysterious
    - C) Angry and confrontational
    - D) Reflective and nostalgic
4. **What can be inferred about the stranger from his interaction with the innkeeper?**
    - A) He is rude and demanding.
    - B) He is lost and seeking directions.
    - C) He is weary and in need of rest.
    - D) He is a criminal on the run.
5. **Which of the following best captures the theme of the passage?**
    - A) The unpredictability of life in a small village
    - B) The challenges of travel in unfamiliar places
    - C) The tension between curiosity and caution
    - D) The joy of meeting new people in unexpected circumstances

Passage 2 (History/Social Studies):

Passage Title: *The Impact of the Printing Press on European Society*

The invention of the printing press by Johannes Gutenberg in the mid-15th century revolutionized the way information was disseminated and had profound effects on European society. Before the printing press, books were copied by hand, a laborious and time-consuming process that made books expensive and accessible only to the wealthy elite. The printing press changed that by allowing books to be produced quickly and in large quantities, drastically reducing their cost and making them available to a much wider audience.

One of the most significant impacts of the printing press was the spread of literacy. As books became more affordable and widely available, more people learned to read, and literacy rates across Europe began to rise. This democratization of knowledge empowered individuals and contributed to the development of a more informed and engaged populace.

The printing press also played a crucial role in the spread of new ideas and the acceleration of the Renaissance. Scholars, scientists, and philosophers could now share their discoveries and theories with a broader audience, leading to an unprecedented exchange of ideas. The works of classical authors were rediscovered and widely circulated, fueling the intellectual movement that characterized the Renaissance.

Furthermore, the printing press had a profound impact on religion. The ability to produce large numbers of religious texts, including the Bible, in vernacular languages allowed for greater access to religious teachings. This accessibility contributed to the questioning of established religious authorities and was a significant factor in the Protestant Reformation. Reformers like Martin Luther used the printing press to disseminate their ideas rapidly, challenging the Catholic Church and leading to religious upheaval across Europe.

In summary, the invention of the printing press was a turning point in European history. It not only facilitated the spread of knowledge and ideas but also played a key role in shaping the social, intellectual, and religious landscape of the continent.

Questions:

1. **What was one of the primary effects of the printing press on European society?**
    - A) It increased the cost of books.
    - B) It limited access to information.
    - C) It made books more affordable and widely available.
    - D) It discouraged the spread of new ideas.
2. **How did the printing press contribute to the rise in literacy rates in Europe?**
    - A) By making books cheaper and more accessible to the general public.

- B) By replacing traditional methods of education with printed materials.
    - C) By forcing the wealthy elite to educate the masses.
    - D) By limiting book production to religious texts only.
3. **According to the passage, what role did the printing press play in the Renaissance?**
    - A) It delayed the spread of Renaissance ideas due to the slow production of books.
    - B) It enabled the widespread distribution of classical works, fueling the Renaissance.
    - C) It restricted access to new scientific discoveries to a select few.
    - D) It had little to no impact on the intellectual movement of the Renaissance.
4. **How did the printing press influence the Protestant Reformation?**
    - A) It prevented the spread of Reformation ideas by controlling book production.
    - B) It allowed reformers to quickly disseminate their ideas and challenge the Catholic Church.
    - C) It supported the Catholic Church's efforts to suppress heretical ideas.
    - D) It reduced the number of people who could access religious texts.
5. **Which of the following best describes the overall impact of the printing press on European society?**
    - A) It reinforced the existing social hierarchy by restricting knowledge to the elite.
    - B) It transformed European society by democratizing knowledge and facilitating social change.
    - C) It had minimal impact, as most people continued to rely on oral traditions.
    - D) It led to the decline of religious influence in European life.

Passage 3 (Humanities):

Passage Title: *The Influence of Greek Philosophy on Modern Thought*

The philosophical ideas developed in ancient Greece have had a lasting impact on modern thought and continue to influence contemporary discussions in various fields, including ethics, politics, science, and metaphysics. The work of philosophers such as Socrates, Plato, and Aristotle laid the groundwork for Western philosophy and provided a framework that has been built upon by countless thinkers throughout history.

Socrates, often considered the father of Western philosophy, introduced a method of inquiry that involved asking probing questions to stimulate critical thinking and illuminate ideas. This method, known as the Socratic method, is still widely used in education today, particularly in law schools, where it is employed to encourage students to explore complex issues from multiple perspectives.

Plato, a student of Socrates, further developed his teacher's ideas and introduced the concept of ideal forms, which he believed represented the true reality behind the physical world. Plato's theory of forms has influenced a wide range of philosophical discussions, particularly in the realms of metaphysics and epistemology. His work also laid the foundation for political philosophy, as seen in his famous work "The Republic," where he explores the nature of justice and the ideal state.

Aristotle, a student of Plato, took a more empirical approach to philosophy, emphasizing observation and experience as the basis for knowledge. His contributions to logic, biology, ethics, and politics have had a profound impact on the development of these fields. Aristotle's ethical theories, particularly his concept of the "golden mean," which advocates for moderation and balance in all things, continue to be relevant in contemporary discussions on moral philosophy.

The influence of Greek philosophy is also evident in modern science. The scientific method, which involves forming hypotheses, conducting experiments, and drawing conclusions based on empirical evidence, has its roots in the logical and observational approaches pioneered by Aristotle. Additionally, the ideas of Greek philosophers have shaped the development of democratic ideals, human rights, and the concept of the rule of law, all of which are fundamental to modern political systems.

In conclusion, the philosophical contributions of ancient Greece have had an enduring impact on modern thought. The ideas of Socrates, Plato, and Aristotle continue to inspire and inform contemporary discussions in a wide range of disciplines, demonstrating the timeless relevance of their work.

Questions:

1. **What method introduced by Socrates is still widely used in education today?**
   - A) The method of rote memorization
   - B) The Socratic method, involving probing questions
   - C) The method of lecturing and passive listening
   - D) The method of experimental observation
2. **According to the passage, what is Plato's theory of forms?**
   - A) The idea that the physical world is the only reality
   - B) The belief that true reality is found in abstract, ideal forms
   - C) The notion that knowledge comes solely from sensory experience
   - D) The theory that all things are constantly changing and evolving
3. **How did Aristotle's approach to philosophy differ from that of Plato?**
   - A) Aristotle rejected empirical observation in favor of abstract reasoning.
   - B) Aristotle emphasized observation and experience as the basis for knowledge.
   - C) Aristotle focused exclusively on metaphysics, unlike Plato.
   - D) Aristotle believed in ideal forms, whereas Plato did not.
4. **Which of the following best describes Aristotle's concept of the "golden mean"?**
   - A) The pursuit of extreme behaviors for happiness
   - B) The idea that all knowledge is relative
   - C) The belief in moderation and balance in all aspects of life
   - D) The principle that physical pleasure is the highest good
5. **What impact did Greek philosophy have on the development of modern science?**
   - A) It introduced the concept of random experimentation without hypothesis.
   - B) It discouraged the use of empirical evidence in scientific inquiry.
   - C) It laid the foundation for the scientific method based on observation and logic.
   - D) It promoted the idea that science should remain separate from philosophy.

Passage 4 (Science):

Passage Title: *The Role of Wetlands in Environmental Health*

Wetlands, which include marshes, swamps, and bogs, are among the most productive ecosystems on Earth. These areas are characterized by the presence of water, either at or near the surface, for at least part of the year. Wetlands play a crucial role in maintaining environmental health and supporting biodiversity, yet they are often undervalued and face significant threats from human activities.

One of the key functions of wetlands is their ability to act as natural water filters. As water moves through a wetland, it slows down, allowing sediments and pollutants to settle out. This process helps to purify the water before it flows into rivers, lakes, and oceans. Wetlands can also absorb excess nutrients, such as nitrogen and phosphorus, which can otherwise lead to harmful algal blooms in aquatic systems.

In addition to water purification, wetlands provide essential habitat for a wide variety of species. Many species of birds, fish, amphibians, and plants depend on wetlands for their survival. Migratory birds, in particular, rely on wetlands as stopover points during their long journeys, where they can rest and refuel. The loss of wetlands can therefore have a significant impact on global biodiversity.

Wetlands also play a vital role in flood control. Their ability to absorb and store large amounts of water helps to mitigate the effects of heavy rainfall and reduce the risk of flooding in surrounding areas. This function is particularly important in regions prone to seasonal flooding, where wetlands act as natural buffers that protect human settlements and agricultural lands.

Despite their importance, wetlands are among the most threatened ecosystems in the world. Over the past century, many wetlands have been drained, filled in, or converted to other uses, such as agriculture or urban development. This loss of wetlands has not only reduced their ability to provide essential ecological services but has also contributed to the decline of many species that depend on these habitats.

Efforts to protect and restore wetlands are now gaining momentum, as the value of these ecosystems becomes more widely recognized. Conservation initiatives include the creation of protected areas, restoration projects to rehabilitate degraded wetlands, and policies aimed at preventing further loss. By safeguarding wetlands, we can help ensure the continued health of our environment and the diversity of life it supports.

Questions:

1. **What is one of the primary functions of wetlands as described in the passage?**
   - A) They serve as major sources of drinking water.
   - B) They act as natural water filters, purifying water before it flows into larger bodies of water.
   - C) They primarily support the growth of agricultural crops.
   - D) They are used mainly for recreational activities.

2. **How do wetlands contribute to flood control?**
    - A) By accelerating the flow of water to prevent buildup.
    - B) By absorbing and storing large amounts of water, reducing the risk of flooding.
    - C) By diverting water away from human settlements.
    - D) By preventing rainfall from reaching the ground.
3. **According to the passage, why are wetlands important for biodiversity?**
    - A) They are the primary producers of oxygen in aquatic environments.
    - B) They provide essential habitats for a wide variety of species.
    - C) They are the only ecosystems where migratory birds can rest.
    - D) They are the largest ecosystems on Earth.
4. **What is a significant threat to wetlands mentioned in the passage?**
    - A) Overpopulation of species within wetlands
    - B) Conversion of wetlands to agricultural or urban areas
    - C) Lack of water flow due to climate change
    - D) Overuse of wetlands for recreational purposes
5. **What efforts are being made to protect wetlands, according to the passage?**
    - A) Construction of dams to regulate water flow
    - B) Initiatives to restore degraded wetlands and prevent further loss
    - C) Programs to increase agricultural use of wetland areas
    - D) Development of wetlands for urban housing projects

Passage 5 (Editing):

Passage Title: *The Role of Nutrition in Cognitive Health*

Proper nutrition is essential for maintaining not only physical health but also cognitive function. Numerous studies have shown that what we eat can have a profound impact on our brain health, affecting everything from memory and concentration to mood and overall mental well-being. However, the modern diet, which is often high in processed foods and low in essential nutrients, may be contributing to a decline in cognitive health across populations.

Original Passage:

The brain requires a variety of nutrients to function optimally. Omega-3 fatty acids, which are found in fish, nuts, and seeds, are particularly important for brain health. These healthy fats have been shown to improve memory and cognitive performance. Additionally, antioxidants, such as vitamins C and E, help protect the brain from oxidative stress, which can lead to cognitive decline.

Questions:

1. **Which sentence in the passage contains an error in subject-verb agreement?**
    - A) "The brain requires a variety of nutrients to function optimally."
    - B) "Omega-3 fatty acids, which are found in fish, nuts, and seeds, is particularly important for brain health."
    - C) "These healthy fats have been shown to improve memory and cognitive performance."
    - D) "Antioxidants, such as vitamins C and E, help protect the brain from oxidative stress."
2. **Which version of the following sentence best corrects the error in parallel structure?**
    - **Original Sentence:** "The modern diet is high in processed foods and low in essential nutrients, which contribute to poor physical health and can affect cognitive function."
    - A) The modern diet is high in processed foods and low in essential nutrients, both of which contribute to poor physical health and cognitive function.
    - B) The modern diet is high in processed foods and low in essential nutrients, which contribute to poor physical health and also cognitive function.
    - C) The modern diet is high in processed foods and low in essential nutrients, contributing to poor physical health, and it can affect cognitive function.
    - D) The modern diet is high in processed foods and low in essential nutrients, contributing to poor physical health and affecting cognitive function.
3. **Choose the correct punctuation for the following sentence:**
    - **Original Sentence:** "Studies have shown that diets rich in fruits vegetables whole grains and lean proteins are associated with better cognitive health."
    - A) "Studies have shown that diets rich in fruits, vegetables, whole grains, and lean proteins are associated with better cognitive health."

- B) "Studies have shown that diets rich in fruits vegetables, whole grains, and lean proteins are associated with better cognitive health."
- C) "Studies have shown that diets rich in fruits, vegetables whole grains, and lean proteins, are associated with better cognitive health."
- D) "Studies have shown that diets rich in fruits, vegetables, whole grains, and lean proteins are, associated with better cognitive health."

4. **Which revision improves the clarity of the following sentence?**
   - **Original Sentence:** "In addition to providing physical benefits, diets high in nutrients can also lead to improved brain health, this can include better memory, concentration, and overall cognitive function."
   - A) In addition to providing physical benefits, diets high in nutrients can also lead to improved brain health; this can include better memory, concentration, and overall cognitive function.
   - B) In addition to providing physical benefits, diets high in nutrients can also lead to improved brain health, this can include better memory, concentration and, overall cognitive function.
   - C) In addition to providing physical benefits diets high in nutrients can also lead to improved brain health; this can include better memory, concentration, and overall cognitive function.
   - D) In addition to providing physical benefits, diets high in nutrients can also lead to improved brain health; this includes better memory, concentration, and overall cognitive function.
5. **Identify the most effective way to combine the following sentences:**
   - **Original Sentences:** "Proper nutrition is essential for maintaining cognitive function. The modern diet is often lacking in the nutrients needed for optimal brain health."
   - A) Proper nutrition is essential for maintaining cognitive function, but the modern diet is often lacking in the nutrients needed for optimal brain health.
   - B) Proper nutrition is essential for maintaining cognitive function, and the modern diet is often lacking in the nutrients needed for optimal brain health.
   - C) Proper nutrition is essential for maintaining cognitive function because the modern diet is often lacking in the nutrients needed for optimal brain health.
   - D) Proper nutrition is essential for maintaining cognitive function; the modern diet is often lacking in the nutrients needed for optimal brain health.

KEY ANSWERS

	Q1	Q2	Q3	Q4	Q5
Passage 1 (Module 1)	B	D	B	C	B
Passage 2 (Module 1)	C	A	C	B	C
Passage 3 (Module 1)	C	B	C	B	C

Passage 4 (Module 1)	B	B	C	C	A
Passage 5 (Module 1)	B	A	C	B	B
Passage 1 (Module 2)	B	C	B	C	C
Passage 2 (Module 2)	C	A	B	B	B
Passage 3 (Module 2)	B	B	B	C	C
Passage 4 (Module 2)	B	B	B	B	B
Passage 5 (Module 2)	B	D	A	A	A

Section 2: Math

Total Time: 70 minutes

Total Questions: 44

Module 1: Math

Time: 35 minutes

Total Questions: 22

Question 1:

Solve for x *in the equation:* $3x - 7 = 2x + 5$

- A) 12
- B) 7
- C) 5
- D) -12

Question 2:

If the function $f(x) = 2x^2 - 3x + 4$, *what is the value of* $f(2)$?

- A) 6
- B) 10
- C) 12
- D) 14

Question 3:

The sum of three consecutive integers is 51. What is the smallest of these integers?

- A) 16
- B) 17
- C) 18
- D) 19

Question 4:

If $x^2 - 4x - 5 = 0$, *what are the solutions for* x*?*

- A) x = 5 or x=-1
- B) x=4 or x=1
- C) x=3 or x=3
- D) x=2 or x=2

Question 5:

In the figure below, if the lengths of the sides of a right triangle are 3, 4, and x, what is the length of the hypotenuse?

- A) 3
- B) 4
- C) 5
- D) 7

Question 6 (SPR):

If $5y + 7 = 2y + 16$, *what is the value of y?*

Question 7:

What is the equation of the line that passes through the points (1,2) and (3,6)?

- A) $y = 2x + 1$
- B) $y = 2x$
- C) $y = 2x + 2$
- D) $y = 4x + 1$

Question 8:

The perimeter of a rectangle is 40 units. If the length is twice the width, what is the area of the rectangle?

- A) 20
- B) 40
- C) 60
- D) 80

Question 9 (SPR):

What is the value of $2^3 + 4^2$*?*

Question 10:

A circle has a radius of 7 units. What is the circumference of the circle? ($Use\ \pi \approx 3.14$)

- A) 21.98
- B) 43.96
- C) 49.84
- D) 56.52

Question 11:

Which of the following is the solution to the inequality $3x - 5 < 7$*?*

- A) x>2
- B) x>4
- C) x<2
- D) x<4

Question 12:

What is the slope of the line represented by the equation $4x - 2y = 8$?

- A) 2
- B) -2
- C) 1/2
- D) -1/2

Question 13 (SPR):

If the area of a square is 49 square units, what is the length of one side?

Question 14:

The sum of the interior angles of a polygon with n sides is given by the formula $180(n - 2)$. If the sum of the interior angles of a polygon is 720 degrees, how many sides does the polygon have?

- A) 4
- B) 5
- C) 6
- D) 8

Question 15:

The probability of drawing a red card from a standard deck of 52 cards is closest to which of the following?

- A) $\frac{1}{2}$
- B) $\frac{1}{4}$
- C) $\frac{1}{3}$
- D) $\frac{2}{3}$

Question 16 (SPR):

Solve for z in the equation $4z + 9 = 25$.

Question 17:

If $f(x) = 2x + 1$, what is the value of $f^{-1}(x)$?

- A) $\frac{x-1}{2}$
- B) $2x - 1$
- C) $\frac{x+1}{2}$
- D) $\frac{x}{2} - 1$

Question 18:

In a geometric sequence, the first term is 2, and the common ratio is 3. What is the fourth term?

- A) 18
- B) 54
- C) 162
- D) 486

Question 19 (SPR):

If $x^2 + 4x = 12$, *solve for* x.

Question 20:

If $\sin\theta = \frac{1}{2}$ *what is* θ*?*

- A) 30°
- B) 45°
- C) 60°
- D) 90°

Question 21:

What is the standard deviation of the data set 2,4,4,4,5,5,7,92, 4, 4, 4, 5, 5, 7, 92,4,4,4,5,5,7,9?

- A) 1
- B) 2
- C) 3
- D) 4

Question 22:

Which of the following equations represents a parabola that opens upwards?

- A) $y = -2x^2 + 3x - 4$
- B) $y = 3x^2 - 2x + 1$
- C) $y = -x^2 + 4x - 5$
- D) $y = 2x^2 - 3x + 7$

Module 2: Math

Time: 35 minutes

Total Questions: 22

Question 1:

If $x - 3 = 7$, *what is the value of* x?

- A) 3
- B) 7
- C) 10
- D) 13

Question 2:

Simplify the expression $4(x - 2) + 3(2x + 5)$.

- A) $14x + 7$
- B) $10x - 7$
- C) $10x + 7$
- D) $14x - 7$

Question 3:

The length of a rectangle is twice its width. If the perimeter of the rectangle is 36 units, what is the length of the rectangle?

- A) 6 units
- B) 12 units
- C) 18 units
- D) 24 units

Question 4 (SPR):

What is the value of $5^2 - 3 \times 4$?

Question 5:

If $y = 3x + 2$ *and* $y = 5x - 4$, *at what point do the lines intersect?*

- A) (3,11)
- B) (3,4)
- C) (4,3)
- D) (−3,−4)

Question 6:

What is the solution to the inequality $2x+3>7$?

- A) $x>2$
- B) $x>1$
- C) $x<2$
- D) $x<1$

Question 7:

If the area of a circle is 64π square units, what is the radius of the circle?

- A) 4 units
- B) 8 units
- C) 16 units
- D) 32 units

Question 8 (SPR):

Solve for n in the equation $3n - 4 = 2n + 5$.

Question 9:

Which of the following represents the graph of the equation $y = -x^2 + 4x - 3$?

- A) A parabola opening upwards
- B) A parabola opening downwards
- C) A straight line with a positive slope
- D) A straight line with a negative slope

Question 10:

The average (arithmetic mean) of five numbers is 14. If four of the numbers are 12, 15, 18, and 10, what is the fifth number?

- A) 13
- B) 14
- C) 15
- D) 17

Question 11 (SPR):

If $7x - 3 = 18$, what is the value of x?

Question 12:

What is the slope of the line that passes through the points (2,3) and (5,11)?

- A) 2
- B) $\frac{8}{3}$
- C) 4
- D) $\frac{3}{8}$

Question 13:

Which of the following is equivalent to $\sqrt{50}$*?*

- A) $5\sqrt{2}$
- B) $10\sqrt{5}$
- C) $25\sqrt{2}$
- D) $2\sqrt{5}$

Question 14:

In a triangle, one angle measures 90°, and another angle measures 45°. What is the measure of the third angle?

- A) 30°
- B) 45°
- C) 60°
- D) 90°

Question 15:

Which of the following functions is increasing for all values of x?

- A) $f(x) = -x^2 + 3x + 2$
- B) $f(x) = x^2 - 2x - 1$
- C) $f(x) = 2x - 1$
- D) $f(x) = -3x + 4$

Question 16 (SPR):

Solve for k in the equation $k^2 - 9 = 0$.

Question 17:

If $f(x) = 3x + 1$, *what is the value of f(4)?*

- A) 7
- B) 10
- C) 13
- D) 16

Question 18:

If the lengths of the sides of a triangle are 7, 24, and 25 units, what is the area of the triangle?

- A) 70 square units

- B) 84 square units
- C) 120 square units
- D) 168 square units

Question 19:

What is the distance between the points (1,2) and (4,6) in the coordinate plane?

- A) 3 units
- B) 4 units
- C) 5 units
- D) 6 units

Question 20:

Which of the following is an example of a linear function?

- A) $y = x^2 + 2x + 1$
- B) $y = 3x + 2$
- C) $y = \frac{1}{x} + 4$
- D) $y = \sqrt{x + 4}$

Question 21 (SPR):

If $2m + 7 = 19$, *what is the value of* m*?*

Question 22:

The expression (x + 2)(x - 3) is equivalent to which of the following?

- A) $x^2 - x - 6$
- B) $x^2 + x - 6$
- C) $x^2 - 5x - 6$
- D) $x^2 + 5x - 6$

Section 2: Math - Module 1

1. C) 10
2. D) 14
3. B) 17
4. A) x=5 or x=−1
5. C) 5
6. SPR) y=3
7. B) y=2x
8. B) 40
9. SPR) 24
10. B) 43.96
11. B) x>4

12. D) $-1/2$
13. SPR) 7
14. C) 6
15. A) $\frac{1}{2}$
16. SPR) z=4
17. A) $\frac{x-1}{2}$
18. B) 54
19. SPR) x=−6 or x=2
20. A) 30°
21. B) 2
22. B) $y = 3x^2 - 2x + 1$

Section 2: Math - Module 2

1. C) 10
2. C) 10x + 7
3. B) 12 units
4. SPR) 13
5. D) (-3, -4)
6. B) x>2
7. B) 8 units
8. SPR) n=9
9. B) A parabola opening downwards
10. D) 17
11. SPR) x=3
12. C) 4
13. A) $5\sqrt{2}$
14. B) 45°
15. C) $f(x) = 2x - 1$
16. SPR) k=3 or k=−3
17. C) 13
18. B) 84 square units
19. C) 5 units
20. B) $y = 3x + 2$
21. SPR) m=6
22. A) $x^2 - x - 6$

Essay Section

Time Allotted: 50 minutes

Directions:

The following essay prompt presents a brief excerpt from a larger argument. Your task is to read the passage and analyze how the author builds an argument to persuade their audience. In your essay, you should focus on how the author uses reasoning, evidence, and stylistic or persuasive elements to support their claims.

You are not being asked to agree or disagree with the author's position. Instead, you should evaluate the effectiveness of the author's argument by discussing the techniques they use to build it. Your analysis should be precise and well-organized, with a clear focus on the most relevant features of the passage.

Essay Prompt:

"The most effective way to influence the world is to start with oneself. Personal growth and development lay the foundation for broader societal change. When individuals strive to improve their character, skills, and knowledge, they set an example for others and create a ripple effect that can transform communities and nations."

Task:

Write an essay in which you explain how the author builds an argument to persuade the audience of the validity of this statement. Be sure that your analysis focuses on the most relevant features of the passage and how they contribute to the overall persuasiveness of the author's argument.

Guidelines for Writing:

1. **Plan Your Response:** Take a few minutes to read the passage and organize your thoughts. Consider which aspects of the author's argument are most compelling and how you can effectively analyze them.
2. **Structure Your Essay:** Begin with an introduction that clearly states the author's argument and the techniques you will be analyzing. Develop your analysis in the body paragraphs, using specific examples from the passage to support your points. Conclude with a summary of how the author's techniques contribute to the overall persuasiveness of the argument.
3. **Use Clear and Concise Language:** Aim for clarity and precision in your writing. Avoid unnecessary words and focus on making your analysis as effective and straightforward as possible.
4. **Proofread:** If time permits, review your essay for any grammatical errors or unclear sentences. A well-polished essay will help ensure that your analysis is communicated effectively.

Essay Prompt:

Essay Topic:

Consider the following excerpt from a speech:

"The most effective way to influence the world is to start with oneself. Personal growth and development lay the foundation for broader societal change. When individuals strive to improve their character, skills, and knowledge, they set an example for others and create a ripple effect that can transform communities and nations."

Write an essay in which you explain how the author builds an argument to persuade the audience of the validity of this statement. In your essay, analyze how the author uses one or more of the following to strengthen their argument: reasoning, evidence, and stylistic or persuasive elements. Be sure that your analysis focuses on the most relevant features of the passage.

Sample Essay:

In the excerpt, the author argues that personal growth and development are crucial for creating broader societal change. To persuade the audience of this idea, the author employs a combination of logical reasoning, evidence from historical examples, and persuasive language that appeals to the reader's sense of responsibility and potential.

One of the primary strategies the author uses is logical reasoning. The author begins by establishing a clear connection between individual actions and societal outcomes. By stating that "the most effective way to influence the world is to start with oneself," the author suggests that large-scale change begins with small, personal actions. This reasoning is grounded in the idea that societal structures are composed of individuals; therefore, improving oneself inevitably contributes to the improvement of society as a whole. The logic is straightforward: if each person takes responsibility for their growth, the cumulative effect will be a positive transformation of the broader community.

In addition to logical reasoning, the author supports their argument with evidence from historical examples. Although the specific examples are not detailed in the excerpt, the reference to "setting an example for others" implies that history has witnessed individuals whose personal growth led to significant societal changes. Figures like Mahatma Gandhi or Martin Luther King Jr. exemplify this idea—both leaders focused on personal development and nonviolent principles, which eventually inspired movements that changed the course of history. By invoking these implicit examples, the author reinforces the argument that personal growth can have far-reaching effects.

Moreover, the author uses persuasive language to appeal to the audience's emotions and sense of agency. Words like "ripple effect" and "transform communities and nations" evoke a sense of empowerment and hope. The imagery of a ripple spreading across water conveys the idea that even small actions can have expansive impacts. This choice of words encourages the audience to see their personal growth not as an isolated endeavor but as something that can contribute to a larger, positive change. The use of inclusive language, such as "when individuals strive," also creates a sense of collective responsibility, making the audience feel that they are part of a larger movement toward improvement.

Finally, the author's tone is both motivational and authoritative, which helps to solidify the argument. The confident assertion that "personal growth and development lay the foundation for broader societal change" leaves little room for doubt and urges the audience to take the idea seriously. By presenting the argument in such a definitive manner, the author reinforces the notion that personal development is not merely beneficial but essential for societal progress.

In conclusion, the author effectively builds an argument for the importance of personal growth in driving societal change through the use of logical reasoning, historical evidence, and persuasive language. By connecting individual actions to broader outcomes, the author not only convinces the audience of the validity of the argument but also inspires them to take action in their own lives. This combination of strategies makes the argument compelling and resonant with the audience, encouraging them to contribute to the betterment of society through their personal development.

Essay Prompt 1:

Essay Topic:

"Success is not defined by how much you achieve, but by the challenges you overcome in the process. True success is measured by resilience, determination, and the ability to rise after each setback."

Write an essay in which you explain how the author builds an argument to persuade the audience of the validity of this statement. In your essay, analyze how the author uses reasoning, evidence, and stylistic or persuasive elements to strengthen their argument. Focus on the most relevant features of the passage.

Essay Prompt 2:

Essay Topic:

"Innovation often comes from the willingness to question established norms and take risks. Societies that encourage creativity and experimentation tend to make greater advancements in science, technology, and culture."

Write an essay in which you explain how the author builds an argument to persuade the audience of the validity of this statement. In your essay, analyze how the author uses reasoning, evidence, and stylistic or persuasive elements to strengthen their argument. Focus on the most relevant features of the passage.

Essay Prompt 3:

Essay Topic:

"The strength of a community is not measured by the wealth of its members, but by the connections between them. A society that values compassion, cooperation, and mutual support is more resilient in the face of challenges."

Write an essay in which you explain how the author builds an argument to persuade the audience of the validity of this statement. In your essay, analyze how the author uses reasoning, evidence, and stylistic or persuasive elements to strengthen their argument. Focus on the most relevant features of the passage.

Essay Prompt 4:

Essay Topic:

"Education is the most powerful tool we have for shaping the future. By empowering individuals with knowledge and critical thinking skills, we can create a more informed and engaged society that is capable of addressing the challenges of tomorrow."

Write an essay in which you explain how the author builds an argument to persuade the audience of the validity of this statement. In your essay, analyze how the author uses reasoning, evidence, and stylistic or persuasive elements to strengthen their argument. Focus on the most relevant features of the passage.

Essay Prompt 5:

Essay Topic:

"Technology has the potential to connect people across the globe and break down barriers between cultures. However, it can also lead to isolation and a loss of meaningful human interaction. The impact of technology on society depends on how we choose to use it."

Write an essay in which you explain how the author builds an argument to persuade the audience of the validity of this statement. In your essay, analyze how the author uses reasoning, evidence, and stylistic or persuasive elements to strengthen their argument. Focus on the most relevant features of the passage.

Secret Section: Your Ultimate SAT Success Toolkit

Welcome to the **Secret Section**, an exclusive part of your SAT preparation journey designed to provide you with insider knowledge, expert advice, and strategic tools that will help you excel not only on the SAT but also in the college admissions process. This section is your hidden advantage—a collection of powerful resources that will set you apart from other test-takers and help you achieve your academic goals.

Secret 1: The Insider's SAT Scoring Guide

Understanding the nuances of how the SAT is scored can be the difference between a good score and a great score. This guide will take you behind the scenes of the SAT scoring process, giving you the knowledge you need to maximize your results. Here's what you'll discover:

Understanding the SAT Score Report

When you receive your SAT score report, it's easy to focus solely on the final number. However, to truly understand your performance and plan your next steps, it's essential to dive deeper into the details of how your score is calculated.

- **Raw Scores vs. Scaled Scores:**
  Your SAT score starts as a "raw score," which is simply the number of questions you answered correctly in each section. The SAT does not penalize for incorrect answers, so your raw score is purely additive. However, this raw score is then converted into a "scaled score" that ranges from 200 to 800 per section, with a total possible score of 1600.

Why the Conversion Matters:

The process of converting raw scores to scaled scores involves a statistical process known as equating. This ensures that scores are comparable across different test versions, which might vary slightly in difficulty. Understanding this can help you see how even small improvements in the number of correct answers can lead to significant jumps in your scaled score.

- **Section Breakdown:**
  The SAT is divided into two main sections: Evidence-Based Reading and Writing (EBRW) and Math. Each of these sections is scored on a scale of 200-800. The EBRW section is further divided into Reading and Writing & Language, but these subsections do not receive individual scores; rather, their performance contributes to the overall EBRW score.

Interpreting Your Section Scores:

If you excel in one area but struggle in another, it's crucial to understand how these scores balance out. For example, a high Math score can offset a lower EBRW score, and vice versa. Your goal should be to identify your strengths and use them to boost your overall score, while also working to improve weaker areas.

- **Subscores and Cross-Test Scores:**
  In addition to the main section scores, the SAT provides detailed subscores and cross-test scores that give further insight into your performance.

Subscores:

These are reported on a scale of 1-15 and cover specific skill areas such as:

- o **Command of Evidence:** How well you can use evidence to support your answers.
- o **Words in Context:** Your ability to understand the meaning of words in different contexts.
- o **Expression of Ideas and Standard English Conventions:** These are particularly relevant in the Writing and Language section, focusing on your ability to revise and edit text for clarity and correctness.

Cross-Test Scores:

These scores, ranging from 10-40, reflect your performance on questions that test skills across multiple subjects, such as:

- o **Analysis in History/Social Studies:** How well you can analyze historical and social science contexts, which appear in both Reading and Writing sections.
- o **Analysis in Science:** Your ability to interpret and analyze scientific data, relevant in the Reading, Writing, and Math sections.

Leveraging Subscores and Cross-Test Scores:

Colleges may look at these scores to gain a more nuanced understanding of your abilities. High subscores in areas like Command of Evidence or Words in Context can highlight your readiness for college-level work in specific disciplines.

- **Score Reporting and Percentiles:**
  Your SAT score report also includes percentile rankings, which compare your performance to that of other test-takers. For example, if you're in the 75th percentile, you scored higher than 75% of students who took the SAT.

Why Percentiles Matter:

Percentiles give you a sense of where you stand relative to other college-bound students. They can also help you understand how competitive your score is for different colleges. A score that places you in the top 10% might make you a strong candidate for highly selective schools, while a score in the 50th percentile might align better with less selective institutions.

Maximizing Your Score

Once you understand how the SAT is scored, you can develop strategies to maximize your results. Here are some advanced techniques to help you get the most out of your SAT performance:

- **Strategic Guessing:** Since the SAT does not penalize for incorrect answers, guessing on questions where you're unsure can be a beneficial strategy. However, guessing should be done strategically.

When to Guess:

  - **Eliminate Wrong Answers:** Before guessing, try to eliminate as many incorrect options as possible. If you can eliminate one or two answers, your chances of guessing correctly improve significantly.
  - **Time Management:** If you're running out of time, don't leave any questions blank. Quickly fill in answers for any remaining questions, even if you're unsure, to maximize your chances of earning extra points.
- **Focus Areas:** Not all questions on the SAT are created equal in terms of difficulty or point value, but improving your performance on certain types of questions can lead to significant score gains.

High-Impact Areas:

  - **Math Grid-In Questions:** These questions do not have multiple-choice answers and require you to generate the answer on your own. They often contribute more heavily to your raw score, so focusing on accuracy here can have a big impact.
  - **Paired Passages in Reading:** These often involve more complex analysis and synthesis, so mastering them can boost your reading score.
  - **Grammar Rules in Writing:** The Writing & Language section is heavily based on a set of core grammar rules. Mastering these rules can lead to quick and significant improvements.

- **Test Day Strategy:** Preparing for the SAT is not just about knowing the material—it's also about having a clear strategy for test day.

Pacing:

  - **Use Your Time Wisely:** Know the time limits for each section and practice pacing yourself during practice tests. Remember, it's better to move on from a question that's taking too long and come back to it if time allows.
  - **Section Management:** Start with the questions you find easiest. This will build your confidence and ensure that you secure as many points as possible before tackling more difficult questions.

Mindset:

  - **Stay Calm and Focused:** The SAT is a long and demanding test. Practice relaxation techniques, such as deep breathing, to stay calm and focused during the exam.
  - **Positive Visualization:** Before the test, visualize yourself succeeding. Confidence can significantly impact your performance, so enter the exam with a positive mindset.

By mastering the insights and strategies in this Insider's SAT Scoring Guide, you'll be equipped to take full control of your SAT performance. Understanding how your score is calculated, how to optimize it, and how to strategically approach the test will give you a significant advantage on test day.

Secret 3: Mastering Superscoring - Your Path to the Highest Possible SAT Score

Superscoring is a powerful strategy that can significantly enhance your SAT performance and boost your college application. By understanding and effectively using superscoring, you can present your best possible SAT score to colleges, giving you a competitive edge in the admissions process. This deep dive will explore what superscoring is, how it works, and how you can use it to maximize your score.

What is Superscoring?

Superscoring is a policy used by many colleges and universities that allows you to combine your highest section scores from multiple SAT test dates to create your best possible composite score. Rather than considering the scores from a single test date, colleges that practice superscoring will look at your highest scores in each section—Evidence-Based Reading and Writing (EBRW) and Math—regardless of when you took the test.

How Superscoring Works:

- **Multiple Test Dates, Best Section Scores:** Imagine you took the SAT twice. On your first attempt, you scored 680 in EBRW and 650 in Math. On your second attempt, you scored 640 in EBRW and 700 in Math. Without superscoring, your highest total score would be 1380 from your first test date (680 + 700). However, with superscoring, colleges can combine your highest EBRW score (680) from the first test date with your highest Math score (700) from the second test date, giving you a superscored total of 1380 (680 + 700).
- **The Impact of Superscoring:** By using superscoring, you present a composite score that reflects your best performance in each section, potentially increasing your overall score significantly. This higher score can make you a more competitive applicant, especially at schools that place a strong emphasis on standardized test scores.

The Benefits of Superscoring

Superscoring offers several key benefits that can enhance your college application:

- **Maximizes Your Strengths:** Superscoring allows you to showcase your strengths by focusing on the sections where you performed best across different test dates. This can be particularly advantageous if you excelled in different sections on different test dates.

- **Reduces Pressure:** Knowing that you can improve your overall score by performing well in different sections on different test dates can reduce the pressure to achieve a perfect score in one sitting. This can help you approach each test date with a more relaxed mindset, potentially leading to better performance.
- **Strategic Test-Taking:** Superscoring encourages strategic test-taking. You can focus on improving specific sections in each test without worrying about your performance in the other sections. For example, after securing a strong EBRW score, you can dedicate more time and effort to improving your Math score on subsequent test dates.

How to Plan for Superscoring

To fully leverage superscoring, you need to plan your SAT test dates and study strategies carefully. Here's a step-by-step guide to help you make the most of superscoring:

1. Take the SAT Early and More Than Once:

- **Start Early:** Aim to take your first SAT during your junior year of high school, or even earlier if possible. This gives you ample time to retake the test if needed and to benefit from superscoring.
- **Multiple Attempts:** Plan to take the SAT at least two or three times. This increases your chances of achieving high scores in different sections across different test dates, which can then be combined into a higher superscore.

2. Focus on Sectional Improvement:

- **Analyze Your Performance:** After each test, carefully review your score report to identify which sections need improvement. Focus your study efforts on these areas for your next test date.
- **Targeted Practice:** If your Math score was lower than your EBRW score on your first attempt, dedicate more study time to Math before your next test. Conversely, if your EBRW score was lower, concentrate on reading and writing skills.

3. Use Practice Tests to Simulate Real Conditions:

- **Simulate Superscoring:** Take full-length practice tests and use the results to simulate superscoring. Combine your best section scores from different practice tests to see what your superscored SAT result could look like.
- **Timed Sections:** Practice under timed conditions to mimic the pressure of the actual SAT. Focus on pacing yourself to ensure that you can complete each section within the allotted time.

4. Register for Tests Strategically:

- **Space Out Test Dates:** Give yourself enough time between test dates to study and improve. For example, if you take the SAT in October, plan to retake it in December or the following March.
- **Monitor Application Deadlines:** Make sure your final SAT test date aligns with your college application deadlines. If your schools offer superscoring, you'll want to ensure all your scores are received in time to be considered.

Which Colleges Practice Superscoring?

Not all colleges and universities practice superscoring, so it's important to research the policies of the schools to which you're applying. Here's how to find out if a college superscores:

- **Check the College's Admissions Website:** Most colleges clearly state their SAT score policies on their admissions pages. Look for information about superscoring or call the admissions office if the policy isn't clear.
- **Use College Search Tools:** Some college search websites and tools allow you to filter schools based on their superscoring policies. Use these tools to create a list of potential colleges that offer superscoring.
- **Consider Test-Optional Schools:** Even if a school is test-optional, they may still consider superscored SAT results if you choose to submit them. If your superscore is strong, it's often beneficial to include it in your application.

Common Questions About Superscoring

To ensure you fully understand superscoring and how to make it work for you, here are answers to some common questions:

- **Can I combine scores from different years?** Yes, most colleges that practice superscoring will allow you to combine section scores from SATs taken in different years. However, it's important to confirm this with each college.
- **Do all schools superscore the SAT?** No, not all schools practice superscoring. Some colleges consider only your highest composite score from a single test date, while others may look at your highest score from each section across multiple test dates.
- **How do I send my superscored results to colleges?** When sending your SAT scores through the College Board, you can choose which test dates to send. Colleges that superscore will automatically consider your highest section scores. Be sure to send all relevant test dates to ensure they can create your superscore.
- **Is superscoring the same as Score Choice?** No, superscoring and Score Choice are different. Score Choice allows you to choose which SAT scores to send to colleges. Superscoring, on the other hand, involves combining the highest section scores across multiple test dates. You can use Score Choice to send only the test dates that contribute to your best superscore.

Superscoring is a strategic tool that can help you present your best possible SAT score to colleges. By planning your test dates carefully, focusing on sectional improvement, and understanding the policies of the schools you're applying to, you can use superscoring to your advantage. This approach not only increases your chances of admission but also enhances your eligibility for merit-based scholarships and other opportunities.

Secret 2: The College Admissions Blueprint

Your SAT score is a vital part of your college application, but it's just one piece of the puzzle. To truly stand out in the competitive world of college admissions, you need a strategic approach that leverages your SAT score alongside other key elements of your application. In this section, we'll explore how to use your SAT results to enhance your college applications, secure scholarships, and navigate the admissions process with confidence.

How SAT Scores Impact Admissions

The SAT plays a crucial role in the college admissions process, but how much it matters can vary from one institution to another. Here's a detailed look at how your SAT score factors into your overall application and what you can do to make it work in your favor.

- **The Holistic Review Process:** Many colleges and universities use a holistic review process when evaluating applicants. This means they consider your entire application, not just your SAT score. Admissions officers look at your GPA, extracurricular activities, essays, letters of recommendation, and other factors to get a comprehensive picture of who you are as a student and a person.

Understanding Your SAT's Role:

  - **GPA vs. SAT:** For some schools, your GPA may carry more weight than your SAT score, especially if your high school offers a rigorous curriculum. However, your SAT score can still be a critical factor, particularly at more selective institutions.
  - **Test-Optional Policies:** Some colleges have adopted test-optional policies, meaning you can choose whether to submit your SAT score. If your SAT score is strong, it's usually in your best interest to submit it, as it can bolster your application.

- **Targeting Your Dream Schools:** Knowing how to align your SAT score with your college choices is key to creating a realistic and strategic application list. Here's how to approach it:

Categorizing Schools:

  - **Safety Schools:** These are colleges where your SAT score and GPA are above the average admitted student's profile. You have a high likelihood of acceptance at these schools.
  - **Match Schools:** Your SAT score and GPA align closely with the average admitted student's profile. You have a good chance of being accepted, but there is still some competition.
  - **Reach Schools:** These are highly selective institutions where your SAT score and GPA may be below the average admitted student's profile. Admission is less certain, but it's worth applying if you have a strong overall application.

Researching Admissions Requirements:

  - **Score Ranges:** Research the SAT score ranges of admitted students at your target schools. Most colleges publish the middle 50% range of SAT scores, which shows the scores of the middle 50% of admitted students. Aim to be within or above this range for the best chance of acceptance.

- **Superscoring:** Many colleges practice superscoring, where they take your highest section scores across multiple SAT test dates to create your best possible composite score. This can be advantageous if you've taken the SAT more than once and have strong scores in different sections.

Leveraging SAT Scores for Scholarships

In addition to helping you gain admission to your desired colleges, a strong SAT score can also open the door to merit-based scholarships, which can significantly reduce the financial burden of college.

- **Merit-Based Scholarships:** Many colleges and universities offer merit-based scholarships to students with high SAT scores. These scholarships are awarded based on academic achievement rather than financial need, and they can range from a few thousand dollars to full tuition coverage.

Types of Merit-Based Scholarships:

- **Automatic Scholarships:** Some schools automatically consider you for scholarships based on your SAT score when you apply for admission. If your score meets or exceeds certain thresholds, you may receive a scholarship offer without any additional application required.
- **Competitive Scholarships:** Other scholarships are competitive, requiring separate applications, essays, or interviews. These scholarships often have higher SAT score requirements and are awarded to a select group of students.

Maximizing Your Scholarship Opportunities:

- **Research Scholarship Options:** Start by researching the scholarship opportunities at the colleges you're interested in. Many schools have dedicated scholarship pages on their websites where you can find information about available awards and their requirements.
- **Apply Broadly:** Don't limit yourself to just a few scholarships. Apply to as many as you qualify for, including both automatic and competitive awards. The more applications you submit, the better your chances of receiving financial aid.
- **Meet Deadlines:** Scholarship deadlines often differ from regular admissions deadlines. Make sure you know the deadlines for each scholarship and submit your applications on time.

- **National and Private Scholarships:** In addition to college-specific scholarships, there are numerous national and private scholarships that consider SAT scores as part of their criteria. Organizations like the National Merit Scholarship Corporation, the Coca-Cola Scholars Foundation, and others offer substantial awards to high-achieving students.

National Merit Scholarship Program:

- **Qualifying for National Merit:** To qualify for the National Merit Scholarship Program, you need to take the PSAT/NMSQT during your junior year of high school. Your score on this test determines your eligibility for recognition and scholarships.
- **Advancing in the Competition:** High scorers may be named National Merit Semifinalists, and with a strong SAT score, they can advance to Finalist status, making them eligible for scholarships from the National Merit Scholarship Corporation, colleges, and corporate sponsors.

Other Private Scholarships:

- **Corporate and Foundation Scholarships:** Many companies and private foundations offer scholarships based on academic achievement, community service, and leadership. Some of these scholarships require SAT scores as part of the application.

- **Local Scholarships:** Don't overlook local scholarship opportunities from community organizations, businesses, and clubs. These awards may have less competition and can still provide valuable financial assistance.

Creating a Strategic College Application Plan

To maximize your chances of admission and scholarships, you need a well-thought-out application strategy.

Here's how to create a plan that leverages your SAT score and strengthens your overall application:

- **Start Early:** Begin researching colleges, scholarships, and application requirements well before your senior year. The earlier you start, the more time you'll have to refine your SAT score, gather application materials, and complete scholarship applications.
- **Craft a Compelling Personal Narrative:** Your SAT score is important, but it's not the only factor colleges consider. Use your essays, extracurricular activities, and letters of recommendation to tell a compelling story about who you are and what you'll bring to their campus.

Essays:

- **Showcase Your Strengths:** Use your essays to highlight the strengths that your SAT score reflects. If you have a strong reading score, for example, write about your love of literature or your experience in debate.
- **Address Weaknesses:** If one area of your SAT score is weaker, your essays can help mitigate that by showcasing your strengths in other areas. For example, if your math score is lower, you might write about your achievements in a math-related extracurricular activity or how you've overcome challenges in that subject.

- **Letters of Recommendation:**
  - **Choose Wisely:** Select recommenders who know you well and can speak to your strengths, both academically and personally. Their letters should reinforce the positive attributes reflected in your SAT score and other parts of your application.
- **Extracurricular Activities:**
  - **Quality Over Quantity:** Colleges are more interested in the depth of your involvement in extracurricular activities than the number of activities you list. Focus on activities where you've taken on leadership roles or made significant contributions.
  - **Align with Your SAT Strengths:** If you have a strong SAT score in a particular area, such as science or writing, highlight extracurricular activities that align with those strengths, such as participating in science fairs, writing competitions, or school publications.
- **Use Superscoring to Your Advantage:**
  - **Plan Multiple Test Dates:** If your colleges practice superscoring, consider taking the SAT multiple times to maximize your score. Focus on improving different sections in each test to achieve your highest possible composite score.
  - **Report Your Best Scores:** When sending SAT scores to colleges, only send the scores that reflect your best performance. Superscoring policies allow you to combine the best section scores from different test dates, giving you a competitive edge.

By mastering the strategies in this College Admissions Blueprint, you'll be able to craft an application that not only highlights your SAT score but also presents a well-rounded, compelling case for your admission. Whether you're aiming for highly selective schools, securing merit-based scholarships, or both, this blueprint will guide you every step of the way.

Bonus Materials

The Bonus Materials section of your SAT preparation book is designed to provide you with additional tools and resources that will further enhance your study experience and help you achieve your best possible score. These materials are intended to be practical, easy to use, and directly applicable to your daily study routine.

Math Cheat Sheets

In this section, we offer you a powerful study aid: **Math Cheat Sheets** that are essential for your SAT preparation. These cheat sheets provide concise summaries of the key formulas, concepts, and strategies that you need to master the SAT Math sections. They are designed to be quick-reference guides that you can easily consult during your study sessions or in the final days leading up to the test.

What's Included:

- **Key Math Formulas:** All the essential formulas for algebra, geometry, trigonometry, and advanced math topics.
- **Quick Problem-Solving Tips:** Strategies for tackling common math problems quickly and accurately.
- **Important Concepts:** Summaries of fundamental math concepts that are frequently tested on the SAT.

How to Access Your Math Cheat Sheets: These cheat sheets are available as downloadable PDFs, easily accessible via a QR code. Simply scan the QR code below to download the Math Cheat Sheets directly to your device. You can print them out or keep them handy on your phone or tablet for quick review whenever you need.

Make sure to utilize these cheat sheets regularly as part of your study routine—they're your go-to resource for ensuring you have all the critical math knowledge at your fingertips.

Progress Tracking Tools

Tracking your progress is an essential part of your SAT preparation. It allows you to measure how far you've come, identify areas that still need improvement, and adjust your study plan accordingly. The Progress Tracking Tools provided here are designed to help you stay organized, motivated, and on track to reach your target SAT score.

1. Practice Test Score Tracker

Purpose:

The Practice Test Score Tracker is a comprehensive tool for recording your scores from each full-length practice test. It helps you monitor your performance across different sections of the SAT, providing a clear overview of your strengths and areas that need further attention.

How to Use It:

- **Record Your Scores:** After each practice test, log your scores for the Math, Evidence-Based Reading, and Writing & Language sections.
- **Analyze Trends:** Over time, review your scores to identify patterns. Are you consistently improving in one section but not in another? Use this information to adjust your study focus.
- **Set Goals:** Use your recorded scores to set realistic improvement goals for your next practice test. For example, if you scored 600 in Math, aim for a 620 on your next test by focusing on specific areas of weakness.

Format:

- **Date of Test:**
- **Math Score:**
- **Evidence-Based Reading Score:**
- **Writing & Language Score:**
- **Total Score:**
- **Notes:** (Use this section to jot down any observations, such as which types of questions you found most challenging.)

2. Sectional Score Breakdown

Purpose:

The Sectional Score Breakdown tool allows you to dig deeper into your performance within each section of the SAT. By breaking down your scores into specific question types, you can pinpoint exactly where you need to improve.

How to Use It:

- **Detail Your Scores:** After completing each practice test or section, break down your scores by question type. For instance, in the Math section, note how you performed on algebra questions versus geometry or data analysis questions.
- **Identify Weaknesses:** Use this breakdown to identify your weakest areas. For example, if you struggle with geometry questions, allocate more study time to reviewing geometry concepts and practicing related problems.
- **Track Improvement:** Over time, use this tool to track improvements in specific areas, ensuring that your study efforts are paying off.

Format:

- **Section:** (e.g., Math, Reading)
- **Question Type:** (e.g., Algebra, Geometry)
- **Number of Correct Answers:**
- **Number of Incorrect Answers:**
- **Notes:** (Use this section to note any particular difficulties or improvements.)

3. Weekly Progress Reports

Purpose:

Weekly Progress Reports help you keep track of your study habits, ensuring that you're consistently working towards your goals. They encourage accountability and provide a structured way to reflect on your study efforts each week.

How to Use It:

- **Log Study Hours:** Each day, record how much time you spend studying each section of the SAT. At the end of the week, review your total study hours and compare them to your goals.
- **Compare Goals with Achievements:** At the beginning of each week, set specific goals, such as completing a certain number of practice questions or improving your speed in a particular section. At the end of the week, reflect on whether you met these goals and why or why not.
- **Adjust Your Plan:** Use the insights from your weekly report to make necessary adjustments to your study plan. If you're falling behind in one area, allocate more time to it in the coming week.

Format:

- **Week of:**
- **Study Hours Logged (by section):**
- **Goals Set for the Week:**
- **Goals Achieved:**
- **Notes:** (Use this section to reflect on your achievements or any obstacles you encountered.)

4. Strengths and Weaknesses Analysis

Purpose:

The Strengths and Weaknesses Analysis is a self-assessment tool that prompts you to regularly evaluate your performance. This tool helps you focus your efforts where they're needed most.

How to Use It:

- **Evaluate Regularly:** After each practice test or at the end of each week, spend time reflecting on your strengths and weaknesses. What are you consistently doing well? Where do you need more practice?
- **Develop an Action Plan:** Based on your analysis, create a targeted action plan for the following week. Focus on reinforcing your strengths while dedicating extra time to areas of weakness.
- **Monitor Progress:** Over time, use this tool to monitor how your strengths and weaknesses evolve. Adjust your study plan as needed to continue making progress.

Format:

- **Strengths:** (List specific areas where you consistently perform well.)
- **Weaknesses:** (List areas where improvement is needed.)
- **Action Plan for Improvement:** (Outline steps to strengthen your weak areas.)
- **Notes:** (Reflect on your progress and any adjustments to your study strategy.)

Accessing the Progress Tracking Tools: These tools are available as downloadable templates that you can print out or use digitally. Simply scan the QR code below to access the Progress Tracking Tools directly on your device.

Regularly using these tools will keep you organized, help you stay focused, and ensure that you're continuously improving as you prepare for the SAT. Tracking your progress is not just about accountability—it's about seeing your hard work translate into tangible results.

College Admissions Guide

This guide is designed to help you navigate the complex and competitive world of college admissions. It provides insights on how to leverage your SAT scores, craft a compelling application, and maximize your chances of gaining admission to your top-choice schools. Additionally, it offers strategies for securing merit-based scholarships to ease the financial burden of college.

1. Introduction

Overview of the College Admissions Process

The college admissions process is a significant step in your academic journey, often determining where you will spend the next four years of your education. This process involves several components, including your academic record, standardized test scores, extracurricular activities, personal essays, and letters of recommendation. Each of these elements plays a role in shaping your application and influencing the decisions made by admissions committees.

Navigating this process can be overwhelming, but with the right guidance and preparation, you can present a strong application that highlights your strengths and aligns with the expectations of your chosen schools.

The Role of SAT Scores in Admissions

Your SAT score is one of the most critical aspects of your college application. While it's not the only factor, it often serves as a standardized measure of your academic abilities, allowing colleges to compare applicants from diverse educational backgrounds.

SAT scores can impact your application in several ways:

- **Benchmarking Academic Readiness:** Colleges use your SAT score to assess your readiness for college-level work. A high score indicates that you have the skills needed to succeed in rigorous academic environments.
- **Determining Admissions Competitiveness:** Your SAT score is often compared against the scores of other applicants. This comparison helps colleges decide where you stand among the pool of candidates.
- **Influencing Scholarship Decisions:** Many colleges and scholarship programs use SAT scores as a key criterion for awarding merit-based scholarships. A strong SAT score can significantly reduce your college expenses.

This guide will delve into how to strategically use your SAT score in the college admissions process, helping you understand its role and maximize its impact.

2. Understanding How SAT Scores Impact College Admissions

Your SAT score is a crucial component of your college application, but its impact can vary depending on the school and its admissions policies. This section will help you understand how different colleges view SAT scores and how you can strategically use your score to enhance your application.

Holistic Review Process

How Colleges Evaluate Applications Many colleges, especially selective ones, use a holistic review process when evaluating applicants. This means that they consider the entirety of your application rather than focusing solely on your SAT score. While a strong SAT score can boost your application, admissions committees also weigh other factors such as your high school GPA, the rigor of your coursework, extracurricular activities, personal essays, and letters of recommendation.

- **Balanced Application:** A balanced application demonstrates excellence in multiple areas. For example, if your SAT score is strong but your GPA is average, your application can still be competitive if you have outstanding extracurricular achievements or compelling essays.
- **Contextual Evaluation:** Colleges consider your SAT score in the context of your background. They may take into account the resources available at your school, your family's educational background, and any challenges you've overcome. This means that a slightly lower SAT score might be viewed more favorably if it's clear that you've excelled in other ways.

The Balance Between SAT Scores, GPA, and Other Factors While your SAT score is important, it's just one part of the larger picture. Here's how it fits into the overall evaluation:

- **GPA and Class Rank:** Your GPA and class rank often carry significant weight, especially if you've taken advanced or honors courses. A strong academic record can complement a high SAT score, reinforcing your academic abilities.
- **Course Rigor:** Admissions officers look at the difficulty of the courses you've taken. Advanced Placement (AP), International Baccalaureate (IB), and honors courses demonstrate your willingness to challenge yourself academically, which can strengthen your application.
- **Extracurricular Activities:** Your involvement in extracurricular activities, leadership roles, and community service can differentiate you from other applicants with similar academic profiles. These activities showcase your passions, commitment, and time management skills.
- **Essays and Recommendations:** Personal essays provide insight into your character, goals, and what you'll bring to a college community. Letters of recommendation offer an external perspective on your abilities and potential, often highlighting qualities that aren't apparent in test scores or grades.

Score Ranges and Percentiles

Researching Middle 50% SAT Score Ranges for Target Schools

To understand how competitive your SAT score is, it's essential to research the middle 50% SAT score range for admitted students at each college you're considering. The middle 50% range represents the range of scores between the 25th and 75th percentiles of admitted students.

- **What This Range Means:** If your SAT score falls within or above this range, you're likely a competitive applicant for that school. If your score is below this range, you may need to bolster other parts of your application or consider applying to schools where your score is within the middle 50%.
- **How to Find This Information:** Colleges typically publish their middle 50% SAT score ranges on their admissions websites or in their Common Data Set. Use this data to evaluate how your score stacks up against those of previously admitted students.

Interpreting Your Percentile Rank

Your percentile rank on the SAT shows how your score compares to those of other test-takers. For example, if you're in the 75th percentile, you scored higher than 75% of students who took the SAT.

- **Using Percentiles to Gauge Competitiveness:** Percentiles are especially useful when comparing your performance across different sections of the SAT. If your Math score is in the 90th percentile but your Reading score is in the 60th percentile, you might focus your study efforts on improving your Reading score to create a more balanced application.
- **Percentiles and School Selection:** Some colleges may place more emphasis on specific sections of the SAT based on your intended major. For instance, engineering programs might prioritize Math scores, while humanities programs might focus on Reading and Writing. Understanding where you stand in each percentile can guide your college selection and application strategy.

Test-Optional Policies

What Test-Optional Means

In recent years, many colleges have adopted test-optional policies, meaning that applicants can choose whether or not to submit SAT or ACT scores as part of their application. This policy was expanded due to the COVID-19 pandemic, but many schools have continued to offer it.

- **When to Submit SAT Scores:** If your SAT score is strong and falls within or above the middle 50% range for a college, it's usually advantageous to submit your score. Doing so can strengthen your application, especially if your GPA or other aspects of your academic record are slightly weaker.
- **When to Opt-Out:** If your SAT score is below the middle 50% range and you feel that other parts of your application are stronger, you might choose not to submit your score. Instead, you can focus on highlighting your GPA, coursework, essays, and extracurricular achievements.

Impact of Test-Optional Policies

- **Broadening Access:** Test-optional policies aim to increase access to higher education by reducing barriers for students who may not perform well on standardized tests. This can benefit students from underrepresented backgrounds or those who have faced significant challenges.

- **Holistic Evaluation:** Colleges that are test-optional tend to place greater emphasis on other aspects of your application, such as your essays, recommendations, and extracurricular activities. This means that a well-rounded application can be just as competitive as one with a strong test score.

Superscoring

How Superscoring Works Superscoring is a policy where colleges consider your highest section scores across multiple SAT test dates. For example, if you scored 650 in Math on one test date and 700 in Math on another, the college will use the 700 in their evaluation, even if your other section scores were higher on a different date.

- **Maximizing Your Score:** Superscoring allows you to present your best possible composite score by focusing on improving one section at a time. This strategy can be particularly useful if you have uneven strengths across different sections.
- **Planning for Superscoring:** If you know that a college superscores, you can plan to take the SAT multiple times, focusing on different sections during each attempt. This approach can help you achieve the highest possible score that reflects your abilities.

List of Colleges That Practice Superscoring Not all colleges practice superscoring, so it's important to research the policies of the schools on your list. Many selective schools, including Ivy League universities and top liberal arts colleges, do offer superscoring.

- **How to Find This Information:** Check the admissions websites of the colleges you're applying to or contact their admissions offices directly to confirm their superscoring policy.

Strategies for Maximizing Your Superscore

- **Targeted Study:** After your first SAT attempt, analyze your score report to identify which sections need improvement. Then, focus your study efforts on those areas before your next test date.
- **Multiple Test Dates:** Plan to take the SAT at least two or three times, allowing you to focus on improving specific sections in each attempt.
- **Practice Under Real Conditions:** Simulate the test environment when practicing, using full-length practice tests to get used to the timing and pressure of the SAT. This will help you perform your best on each section, ultimately boosting your superscore.

This section provides a comprehensive understanding of how your SAT score impacts college admissions, how to strategically submit your scores, and how to maximize your results through superscoring. With this knowledge, you can make informed decisions that enhance your overall application and increase your chances of admission to your top-choice schools.

3. Leveraging SAT Scores for Scholarships

A strong SAT score doesn't just enhance your college application—it can also open the door to substantial financial aid through scholarships. Leveraging your SAT score effectively can significantly reduce the cost of college, making your education more affordable. This section will guide you through the types of scholarships available, how to find them, and how to apply successfully.

Merit-Based Scholarships

Types of Merit-Based Scholarships Merit-based scholarships are awarded based on your academic achievements, including your SAT scores, rather than financial need. These scholarships can range from a few thousand dollars to full tuition coverage, and they are offered by colleges, universities, and private organizations.

- **Automatic Scholarships:**
  Some colleges automatically consider you for merit-based scholarships when you apply for admission. These scholarships are awarded based on your SAT score and GPA without requiring a separate application. If your scores meet the institution's criteria, you may receive a scholarship offer along with your admission letter.
- **Competitive Scholarships:**
  Competitive scholarships often require a separate application, which may include essays, interviews, and letters of recommendation. These scholarships typically have higher SAT score requirements and are awarded to a select group of students who demonstrate exceptional academic ability and potential.
- **Renewable vs. One-Time Scholarships:**
  Some merit-based scholarships are renewable, meaning you'll receive the scholarship each year you're enrolled in college as long as you maintain certain academic standards. Others are one-time awards that cover only the first year of college. Be sure to understand the terms of each scholarship you're applying for.

Examples of Colleges with Generous Merit Scholarships Many colleges offer generous merit-based scholarships to attract high-achieving students. Here are a few examples:

- **University of Alabama:** Offers several merit-based scholarships, including the Presidential Scholarship, which covers full tuition for students with a high SAT score and GPA.
- **Boston University:** The Trustee Scholarship provides full tuition to a small number of students who demonstrate exceptional academic achievement and leadership.
- **University of Southern California (USC):** Offers the USC Merit Scholarship, which covers half-tuition for students with outstanding academic records and test scores.

National and Private Scholarships

National Merit Scholarship Program The National Merit Scholarship Program is one of the most prestigious scholarship competitions in the United States. It starts with the PSAT/NMSQT, which you typically take in your junior year of high school. If your PSAT score qualifies, you may become a National Merit Semifinalist and later a Finalist, making you eligible for significant scholarships.

- **Steps to Qualify:**
  - **PSAT/NMSQT:** Your journey begins with taking the PSAT, which serves as the qualifying test for the National Merit Scholarship Program. High scorers are recognized as Commended Students or Semifinalists.
  - **Advancing to Finalist:** Semifinalists must submit an application, including an SAT score that confirms their PSAT performance, an essay, and a recommendation. Finalists are then considered for National Merit Scholarships.
- **Scholarships Available:**
  National Merit Finalists are eligible for several types of scholarships:
  - **National Merit $2,500 Scholarship:** A one-time award given to approximately 7,600 Finalists.

- **Corporate-Sponsored Merit Scholarships:** Awards provided by corporations, which may be renewable or one-time grants.
- **College-Sponsored Merit Scholarships:** Many colleges offer scholarships to National Merit Finalists who list them as their first-choice school.

Researching National and Private Scholarship Opportunities In addition to the National Merit Scholarship, there are countless other national and private scholarships available. Many of these are based on academic achievement, leadership, community service, and, of course, SAT scores.

- **Tools for Finding Scholarships:**
  - **Fastweb:** A comprehensive scholarship search engine that matches students with scholarships based on their profile.
  - **College Board Scholarship Search:** A tool provided by the organization behind the SAT, offering a wide range of scholarship opportunities.
  - **Scholarships.com:** Another popular database where you can search for scholarships based on your academic background, interests, and other criteria.
- **Scholarships by Major or Interest:**
  Many scholarships are tailored to specific fields of study or extracurricular interests. For example, there are scholarships for students planning to major in STEM fields, those with a strong background in the arts, or those involved in community service.

Tips for Applying to Multiple Scholarships Applying for scholarships is a competitive process, but following these tips can increase your chances of success:

- **Start Early:**
  Begin researching and applying for scholarships in your junior year of high school, or even earlier. Many scholarships have early deadlines, and starting early gives you ample time to gather materials and craft strong applications.
- **Personalize Your Applications:**
  Tailor your essays and application materials to each scholarship's specific criteria. Highlight your SAT score where relevant, but also emphasize your other strengths, such as leadership, community involvement, or academic interests.
- **Apply Broadly:**
  Don't limit yourself to just a few scholarships. Apply to as many as you qualify for to increase your chances of receiving financial aid. Even small scholarships can add up and make a significant impact on your overall college costs.

Researching Scholarships

Tools and Resources for Finding Scholarships Finding the right scholarships requires research, but there are many tools available to help you streamline the process. Here's how to get started:

- **Scholarship Databases:**
  Use scholarship search engines like Fastweb, Scholarships.com, and the College Board's Scholarship Search to find opportunities that match your profile. These databases allow you to filter scholarships by criteria such as SAT score requirements, field of study, and geographic location.
- **School-Specific Scholarships:**
  Research scholarships offered by the colleges you're applying to. Many institutions have dedicated scholarship pages on their websites where you can find information about available awards, eligibility criteria, and application processes.
- **Local Scholarships:**
  Don't overlook local scholarships offered by community organizations, businesses, and clubs. These

scholarships often have fewer applicants, increasing your chances of winning. Check with your high school's guidance office or local civic organizations to find these opportunities.

Creating a Scholarship Application Timeline Applying for scholarships requires organization and planning.

Here's how to create an effective timeline:

- **Junior Year (Spring):**
  Begin researching scholarship opportunities. Take the PSAT/NMSQT to qualify for the National Merit Scholarship Program. Start drafting essays and gathering recommendation letters.
- **Senior Year (Fall):**
  Finalize your scholarship application materials. Apply to early-deadline scholarships first. Continue to take the SAT if you're aiming to improve your score for scholarship purposes.
- **Senior Year (Winter):**
  Apply to scholarships with deadlines in December through February. Many national scholarships have deadlines during this time.
- **Senior Year (Spring):**
  Apply to any remaining scholarships. Confirm that all your application materials have been received by the scholarship organizations.

Common Application Requirements and How to Prepare While each scholarship may have its own

requirements, some common elements include:

- **Essays:**
  Most scholarships require one or more essays. Focus on crafting compelling narratives that highlight your academic achievements, leadership qualities, and personal experiences. Tailor each essay to the specific scholarship.
- **Letters of Recommendation:**
  Choose recommenders who know you well and can speak to your strengths in a way that aligns with the scholarship's criteria. Provide them with plenty of time to write and submit their letters.
- **Transcripts and Test Scores:**
  Be prepared to submit your high school transcript and SAT scores. Ensure that these documents are requested and sent well before the scholarship deadlines.

This section equips you with the knowledge and strategies to effectively use your SAT score in the scholarship application process. By understanding the different types of scholarships available and how to apply for them, you can significantly reduce the financial burden of college and maximize your educational opportunities.

4. Creating a Strategic College Application Plan

Crafting a strategic college application plan is essential for maximizing your chances of admission to your top-choice schools. This section will guide you through the process of researching and targeting colleges, crafting a compelling application, and staying organized throughout the application process. By following these steps, you can ensure that your application showcases your strengths and aligns with the expectations of the schools you're applying to.

Research and Target Colleges

Creating a Balanced College List One of the first steps in the college application process is creating a balanced list of schools that includes a mix of safety, match, and reach schools. This approach ensures that you have a range of options, from schools where you are highly likely to be admitted to those that may be more challenging to get into.

- **Safety Schools:**
  Safety schools are colleges where your academic credentials (SAT scores, GPA, etc.) exceed the average admitted student's profile. These schools are your "safe bet" and are likely to offer you admission. It's important to include a few safety schools on your list to ensure you have solid options.
- **Match Schools:**
  Match schools are institutions where your academic credentials are in line with the average admitted student's profile. You have a good chance of being admitted to these schools, and they should make up the bulk of your college list.
- **Reach Schools:**
  Reach schools are more selective institutions where your academic credentials may be slightly below the average admitted student's profile. Admission to these schools is less certain, but they are worth applying to if they are a good fit for your goals and aspirations.

Aligning Your SAT Score with Your College Choices Your SAT score plays a critical role in determining which schools are safety, match, or reach options. Here's how to align your score with your college choices:

- **Research Score Ranges:**
  For each college on your list, research the middle 50% SAT score range for admitted students. This information is usually available on the college's admissions website or in their Common Data Set. Aim to apply to schools where your SAT score falls within or above this range.
- **Consider Superscoring:**
  If a college practices superscoring, your application may be more competitive. Consider retaking the SAT to improve your superscore, especially if you are targeting reach schools.
- **Factor in Your Intended Major:**
  Some colleges may place greater emphasis on certain sections of the SAT depending on your intended major. For example, STEM programs may prioritize Math scores, while humanities programs may focus more on Reading and Writing. Tailor your application strategy to highlight your strengths in the relevant sections.

Crafting a Compelling Application

Writing Strong Essays Your personal essays are a critical component of your college application. They provide an opportunity to showcase your personality, experiences, and aspirations beyond your academic achievements. Here's how to craft essays that stand out:

- **Showcasing Your Strengths:**
  Use your essays to highlight your academic and extracurricular achievements, particularly those that align with your intended field of study. If your SAT score reflects strengths in a specific area, such as math or writing, discuss how this interest has shaped your academic journey.
- **Telling Your Story:**
  Your essay should tell a story that reveals who you are as a person. Focus on a few key experiences that have influenced your growth and development. Be authentic and let your voice shine through.
- **Addressing Any Weaknesses:**
  If there are areas of your application that are weaker, such as a lower GPA or a gap in extracurricular

involvement, use your essay to provide context. Explain any challenges you've faced and how you've worked to overcome them.

Highlighting Extracurricular Activities Extracurricular activities are an important part of the holistic review process. They demonstrate your passions, leadership abilities, and commitment to your community. Here's how to effectively highlight them in your application:

- **Quality Over Quantity:**
  Colleges are more interested in the depth of your involvement than the number of activities you list. Focus on a few activities where you've taken on leadership roles or made significant contributions.
- **Demonstrating Leadership and Commitment:**
  Highlight roles where you've demonstrated leadership, whether as a club president, team captain, or organizer of a community service project. Commitment to an activity over several years is also a strong indicator of your dedication and reliability.
- **Aligning Extracurriculars with Academic Interests:**
  Whenever possible, align your extracurricular activities with your academic interests. For example, if you're applying to an engineering program, highlight your involvement in robotics clubs, math competitions, or science fairs.

Choosing Effective Letters of Recommendation Letters of recommendation provide colleges with an external perspective on your academic and personal qualities. They can reinforce the strengths highlighted in your application and offer insights into your character and potential.

- **Selecting Recommenders:**
  Choose recommenders who know you well and can speak to your strengths in a way that complements your application. For academic recommendations, consider teachers who have taught you in subjects related to your intended major. For personal recommendations, select individuals who can attest to your character and leadership abilities.
- **Providing Context:**
  Help your recommenders write strong letters by providing them with context. Share your resume, discuss your academic goals, and remind them of specific achievements or projects you worked on in their class. This will enable them to write a more detailed and personalized letter.

Staying Organized with Deadlines

Managing Application and Scholarship Deadlines Staying organized is crucial to ensuring that you meet all your application and scholarship deadlines. Missing a deadline can result in your application being disqualified, so it's important to keep track of all important dates.

- **Creating a Master Calendar:**
  Use a physical or digital calendar to track all your application and scholarship deadlines. Include due dates for college applications, scholarship applications, financial aid forms (like the FAFSA), and standardized test registration.
- **Setting Internal Deadlines:**
  To avoid last-minute stress, set internal deadlines that are several days or even weeks before the actual due dates. This gives you time to review your materials, request any necessary documents, and make final edits to your essays.

- **Using Tools for Tracking Progress:**
  Consider using a college application management tool or spreadsheet to keep track of your progress. Include columns for each school's application status, required documents, and deadlines. Update this regularly to stay on top of your tasks.

Tools for Tracking Progress and Staying on Schedule Here are some tools that can help you stay organized throughout the application process:

- **Google Calendar:**
  Set reminders and deadlines for each application component. You can also share your calendar with family members or advisors who are helping you with the process.
- **Trello or Asana:**
  Use these project management tools to create boards or lists for each college application. Track tasks such as essay drafts, transcript requests, and recommendation letters.
- **Spreadsheets:**
  A simple spreadsheet can be a powerful tool for tracking your application status. Include columns for deadlines, submission status, and any notes or reminders.

This section has provided you with a strategic framework for crafting a strong college application. By carefully researching and targeting colleges, writing compelling essays, highlighting your extracurricular activities, and staying organized with deadlines, you can present yourself as a well-rounded and competitive applicant.

5. Final Tips and Resources

As you approach the final stages of your college application journey, it's important to keep a few key tips in mind to ensure that your application is as strong as possible. This section offers practical advice to help you avoid common pitfalls, maximize the effectiveness of your application, and access additional resources to guide you through the process.

Common Mistakes to Avoid in College Applications

Even the most prepared students can make mistakes in their college applications. Being aware of these common errors can help you avoid them and present a polished, professional application.

1. Missing Deadlines

- **Impact:** Missing a college application or scholarship deadline can result in your application being disqualified, no matter how strong it is.
- **How to Avoid It:** Keep a detailed calendar of all deadlines and set internal deadlines for completing tasks well in advance. Always aim to submit your application at least a few days before the official deadline to account for any unexpected issues.

2. Submitting Generic Essays

- **Impact:** Generic essays that could be sent to any school fail to demonstrate your genuine interest in a particular college, which can weaken your application.

- **How to Avoid It:** Customize each essay to the specific college you're applying to. Mention specific programs, professors, or campus features that attract you to the school. Show how your goals align with what the college offers.

3. Overloading Your Application with Extracurriculars

- **Impact:** Listing too many extracurricular activities without meaningful engagement can dilute your application and make it seem unfocused.
- **How to Avoid It:** Focus on a few key activities where you've made significant contributions or held leadership roles. Quality and impact are more important than quantity.

4. Ignoring Optional Sections

- **Impact:** Skipping optional sections, such as additional essays or interviews, can make it seem like you're not fully invested in the application process.
- **How to Avoid It:** Treat optional sections as opportunities to further strengthen your application. Use these sections to showcase additional skills, interests, or experiences that weren't covered elsewhere.

5. Not Proofreading

- **Impact:** Typos, grammatical errors, and other mistakes can make your application appear rushed and unprofessional.
- **How to Avoid It:** Always proofread your application materials multiple times. Consider asking a teacher, mentor, or family member to review your essays and other documents for errors or unclear phrasing.

Additional Resources for College Planning

There are numerous resources available to help you through the college application process. These resources can provide guidance, answer questions, and offer support as you navigate your way to college.

1. Websites and Tools

- **College Board:** The College Board's website offers a wealth of information on SAT preparation, college search tools, and scholarship opportunities. Their BigFuture platform is particularly helpful for planning your college application journey.
- **Common App:** The Common Application is used by hundreds of colleges and universities. The website offers detailed instructions on how to complete your application, as well as resources for essay writing and financial aid.
- **Fastweb:** Fastweb is one of the largest online scholarship databases. It helps students find scholarships that match their profile and provides tips for applying successfully.

2. Books and Guides

- **"The Fiske Guide to Colleges":** This comprehensive guide offers in-depth profiles of hundreds of colleges and universities, helping you research schools and find the best fit.
- **"On Writing the College Application Essay" by Harry Bauld:** This book provides valuable insights into crafting a standout college essay, with practical tips and examples.
- **"The Princeton Review's College Admissions 101":** A practical guide that covers all aspects of the college admissions process, from researching schools to preparing for interviews.

3. Guidance Counselors and College Advisors

- **School Guidance Counselors:** Your school's guidance counselor is an invaluable resource for college planning. They can help you identify potential colleges, understand application requirements, and provide letters of recommendation.
- **Private College Advisors:** If you need more personalized assistance, consider working with a private college advisor. These professionals can offer tailored advice and support throughout the application process, though they typically charge a fee for their services.

4. Online Communities

- **College Confidential:** An online forum where students, parents, and educators discuss all aspects of the college admissions process. It's a great place to ask questions, share experiences, and find support.
- **Reddit's r/ApplyingToCollege:** This subreddit is a vibrant community where students share advice, resources, and encouragement. It's a useful space for connecting with peers going through the same process.

Final Checklist Before Submitting Applications

Before you hit the submit button, take a moment to go through this final checklist to ensure your application is complete and polished:

- **Have you met all the application requirements?**
  Double-check that you've included all required documents, such as your transcript, SAT scores, essays, and letters of recommendation.
- **Have you proofread your application materials?**
  Review your essays, resume, and other documents for any spelling, grammar, or formatting errors. Ensure that your writing is clear, concise, and free of jargon.
- **Have you tailored your essays to each college?**
  Make sure that each essay reflects your genuine interest in the specific college you're applying to. Avoid using generic responses that could apply to any school.
- **Have you completed the optional sections?**
  If applicable, take advantage of optional essays, interviews, or additional materials that can strengthen your application.
- **Have you confirmed your application fee payment?**
  Ensure that you've paid the application fee or submitted a fee waiver request if applicable. Keep a record of your payment confirmation.
- **Have you received confirmation of submission?**
  After submitting your application, check your email for confirmation from the college. If you don't receive a confirmation within a few hours, follow up with the admissions office to ensure your application was received.

With these final tips and resources in hand, you're well-equipped to navigate the college application process with confidence. Remember, preparation and attention to detail are key to creating a strong application that stands out to admissions committees. Best of luck on your journey to college!

Online Tests

As more educational resources and testing opportunities move online, it's essential to understand how to effectively take tests in a digital format. This bonus section provides tips and tools for taking online tests, ensuring you are prepared to perform your best, whether it's a practice SAT or a timed quiz in an online course.

1. Preparing for Online Tests

Technical Preparation

- **Ensure a Stable Internet Connection:** Before taking any online test, make sure your internet connection is stable and reliable. Consider using a wired connection if possible, as it tends to be more stable than Wi-Fi.
- **Check Your Equipment:** Make sure your computer, keyboard, mouse, and any other necessary peripherals are in good working condition. Test your microphone and camera if they are required for the exam.
- **Update Your Browser and Software:** Ensure that your web browser and any required software (like Zoom or a specific test platform) are updated to their latest versions. This helps avoid compatibility issues during the test.

Creating an Optimal Test Environment

- **Find a Quiet Space:** Choose a distraction-free environment where you can focus on the test. Inform family members or roommates about your testing time to minimize interruptions.
- **Use Proper Lighting:** Ensure your workspace is well-lit. Natural light is ideal, but if that's not possible, make sure your room is evenly lit so that your screen is clearly visible without causing eye strain.
- **Organize Your Workspace:** Keep your workspace clean and clutter-free. Have all the necessary materials, like scratch paper, pens, and a calculator (if allowed), within easy reach.

2. Strategies for Online Test-Taking

Time Management

- **Familiarize Yourself with the Platform:** Before the test, spend time exploring the testing platform. Understand how to navigate between questions, how to mark questions for review, and how to submit your answers.
- **Use a Timer:** Keep track of time using the platform's built-in timer or a separate device. Allocate your time based on the number of questions, leaving some buffer time for review.
- **Prioritize Questions:** Start with the questions you find easiest to build confidence and save more time for difficult ones. Use the flagging feature (if available) to mark questions you want to revisit later.

Maintaining Focus

- **Stay Calm:** Online tests can be stressful, especially if technical issues arise. Practice deep breathing techniques to stay calm and focused.
- **Take Scheduled Breaks:** If the test allows, take short breaks to stretch and rest your eyes. This helps maintain your focus and reduces fatigue during longer tests.
- **Avoid Multitasking:** Resist the temptation to check other tabs or devices during the test. Focus solely on the test to avoid distractions and mistakes.

3. Post-Test Review

Review Your Performance

- **Check Your Answers:** If the test allows you to review your answers before submission, take the time to do so. Double-check for any mistakes or missed questions.
- **Analyze Your Results:** After completing the test, review your results to understand your strengths and areas for improvement. Many online platforms provide detailed feedback that can guide your future study sessions.
- **Reflect on the Experience:** Consider what went well and what could be improved for your next online test. This reflection will help you refine your strategies and be better prepared next time.

4. Accessing Online Test Resources

To help you prepare for and succeed in online tests, we've compiled a set of digital resources, including practice tests, technical guides, and time management tools. These resources are available for download by scanning the QR code below.

What's Included:

- **Practice Tests:** Full-length practice tests that you can take online to simulate the actual test environment

Conclusion: Your Journey to SAT Success and Beyond

As you reach the end of this SAT preparation guide, it's essential to reflect on the journey you've undertaken and the tools and strategies you've acquired. Preparing for the SAT is more than just learning formulas, vocabulary, or test-taking strategies; it's about building the confidence, discipline, and resilience that will serve you well not only on test day but throughout your academic and professional life.

Reflecting on Your SAT Journey

From the moment you decided to take the SAT, you embarked on a path that required dedication and effort. This guide has provided you with a comprehensive roadmap—from diagnostic tests and detailed content reviews to full-length practice tests and strategies for test day. Each section of this book was designed to equip you with the knowledge and skills necessary to perform at your best.

Embracing the Learning Process

Preparing for the SAT is a learning process that involves both your intellect and your mindset. You've encountered challenging problems, learned new concepts, and, at times, struggled with difficult questions. These experiences are all part of the journey. Every problem solved, every passage analyzed, and every essay written contributes to your growth as a student and a thinker.

Building Confidence

Confidence comes from preparation and practice. The more you familiarize yourself with the SAT format, the more confident you'll feel on test day. Through repeated practice and review, you've strengthened your test-taking skills and built the mental stamina needed to excel in a timed, high-pressure environment. Remember that the confidence you've gained through this process is one of your greatest assets.

Developing Resilience

The SAT is a test of resilience as much as it is of academic knowledge. There may have been moments when you felt overwhelmed or discouraged, especially after encountering a tough practice test or a concept that seemed elusive. But each time you pushed through, reviewed your mistakes, and tried again, you built the resilience that will help you not just on the SAT but in any challenging situation you face in the future.

The Broader Impact of SAT Preparation

While this guide has focused on preparing you for the SAT, the skills you've developed extend far beyond the test itself. The discipline, critical thinking, and problem-solving abilities you've honed are transferable to many aspects of your life.

Academic Success

The content knowledge you've gained—whether in math, reading, or writing—will continue to serve you in your high school classes and future college courses. The study habits and time management skills you've developed are essential for academic success in any field of study. As you move forward in your education, you'll find that the rigor and structure of your SAT preparation have laid a strong foundation for tackling more advanced coursework.

College Readiness

Colleges seek students who are not only academically capable but also prepared for the challenges of higher education. Your SAT preparation has helped you become more organized, disciplined, and capable of handling complex tasks under pressure. These are qualities that will make you a successful college student, ready to engage with challenging material, participate in meaningful discussions, and contribute to your campus community.

Lifelong Learning

The process of preparing for the SAT has also reinforced the importance of lifelong learning. Whether you're studying for a test, pursuing a new interest, or learning a new skill, the habits you've cultivated—such as setting goals, practicing regularly, and seeking feedback—are essential for continued personal and professional growth. Embrace the idea that learning doesn't end when you leave the classroom or finish a test; it's a lifelong journey.

Looking Ahead: Beyond the SAT

As you look ahead to your college years and beyond, it's important to recognize that the SAT is just one step on your educational journey. While achieving a high score can open doors to college admissions and scholarships, it's your continued dedication to learning and self-improvement that will determine your long-term success.

Transitioning to College Life

College will present new challenges and opportunities. The skills you've developed during your SAT preparation—critical thinking, time management, resilience—will be crucial as you navigate the academic and social aspects of college life. Whether you're choosing a major, engaging in research, or balancing coursework with extracurricular activities, the habits you've built will help you manage your responsibilities and make the most of your college experience.

Exploring New Opportunities

College is a time for exploration and growth. Take advantage of the opportunities available to you—join clubs, attend lectures, participate in internships, and build relationships with professors and peers. The analytical skills you've developed in your SAT preparation will serve you well in any academic or professional endeavor you choose to pursue. Keep an open mind, and don't be afraid to explore subjects or fields that are new to you.

Setting Long-Term Goals

As you move forward, it's important to set long-term goals that align with your values and aspirations. Whether your goal is to excel in a particular academic discipline, pursue a specific career, or make a meaningful impact in your community, the strategic thinking and goal-setting skills you've developed during your SAT preparation will help you chart a clear path toward achieving your dreams.

Final Thoughts: Celebrating Your Achievement

Preparing for the SAT is a significant achievement in itself. It requires time, effort, and perseverance, and you should take pride in the progress you've made. Whether you've already taken the SAT or are preparing to do so, take a moment to acknowledge your hard work and dedication.

Reflect on Your Growth

Look back on where you started and how far you've come. Reflect on the challenges you've overcome, the knowledge you've gained, and the skills you've developed. Recognize that the effort you've invested in preparing for the SAT has not only prepared you for the test but also equipped you with tools that will benefit you in many areas of your life.

Express Gratitude

Consider the support you've received along the way—whether from teachers, family members, friends, or mentors. Expressing gratitude to those who have helped you on this journey can deepen your sense of accomplishment and strengthen your relationships with the people who care about your success.

Look Forward with Confidence

As you move forward, carry the confidence you've gained with you. Trust in your abilities, and remember that you are capable of achieving great things. The SAT is just one milestone in your journey, and many more opportunities for growth and success await you.

In conclusion, the journey to SAT success is not just about achieving a high score; it's about developing the qualities and skills that will serve you throughout your life. As you continue on your educational path, keep striving for excellence, stay curious, and never stop learning. Your hard work has laid the foundation for a bright and promising future—embrace it with confidence and enthusiasm.

Made in the USA
Las Vegas, NV
04 November 2024